Essentials of Botanical Extraction

Principles and Applications

Essentials of Botanical Extraction

Principles and Applications

Dr. Subhash C. Mandal, M. Pharm., Ph.D

(Former Endeavour Research Fellow, Government of Australia)
(Former Distinguished Education & Research Awardee, AAiPS, USA)
(Former Talented Scientist Awardee, University of Colombo, Sri Lanka)
(Former SAARC Fellow, University Grants Commission, Bangladesh)
(University Grants Commission Research Awardee, India)

Associate Professor
Division of Pharmacognosy
Department of Pharmaceutical Technology
Jadavpur University
Kolkata 700032
India

Dr. Vivekananda Mandal, M. Pharm., Ph.D

(Former WERC Researcher, Government of Japan)

Assistant Professor
Division of Pharmacognosy
Institute of Pharmaceutical Sciences
Guru Ghasidas University (Central University)
Bilaspur 495009
Chhattisgarh
India

Dr. Anup Kumar Das, M. Pharm., Ph.D

Research Scientist
Pavan Structurals Private Limited
R-666, TTC Industrial area, MIDC,
Rabale, Navi Mumbai 400701
Maharashtra
India

AMSTERDAM • BOSTON • HEIDELBERG • LONDON
NEW YORK • OXFORD • PARIS • SAN DIEGO
SAN FRANCISCO • SINGAPORE • SYDNEY • TOKYO

Academic Press is an imprint of Elsevier

Academic Press is an imprint of Elsevier
32 Jamestown Road, London NW1 7BY, UK
525 B Street, Suite 1800, San Diego, CA 92101-4495, USA
225 Wyman Street, Waltham, MA 02451, USA
The Boulevard, Langford Lane, Kidlington, Oxford OX5 1GB, UK

Notices
Knowledge and best practice in this field are constantly changing. As new research
and experience broaden our understanding, changes in research methods, professional
practices, or medical treatment may become necessary.

Practitioners and researchers must always rely on their own experience and knowledge
in evaluating and using any information, methods, compounds, or experiments
described herein. In using such information or methods they should be mindful of
their own safety and the safety of others, including parties for whom they have a
professional responsibility.

To the fullest extent of the law, neither the Publisher nor the authors, contributors, or
editors, assume any liability for any injury and/or damage to persons or property as a
matter of products liability, negligence or otherwise, or from any use or operation of
any methods, products, instructions, or ideas contained in the material herein.

ISBN: 978-0-12-802325-9

British Library Cataloguing in Publication Data
A catalogue record for this book is available from the British Library

Library of Congress Cataloging-in-Publication Data
A catalog record for this book is available from the Library of Congress

For information on all Academic Press publication
visit our website at http://store.elsevier.com/

Printed and bound in the USA

Working together
to grow libraries in
developing countries

www.elsevier.com • www.bookaid.org

Contents

5. Extraction of Botanicals

6.   Classification of Extraction Methods

7.   Innovative Extraction Process Design and Optimization Using Design of Experimental Approach

8. Identification Strategies of Phytocompounds

9. Qualitative Phytochemical Screening

10. Profiling Crude Extracts for Rapid Identification of Bioactive Compounds

Foreword by Sarker

I am absolutely delighted to learn that Dr Subhash C. Mandal, a distinguished scientist in the area of Pharmacognosy and Phytotherapy, and two other learned coauthors from his team have presented us with a useful reference book that encompasses various principles and applications associated with botanical extraction.

As the title, **Essentials of Botanical Extraction – Principles and Applications**, implies, this book integrates insights regarding several conventional and modern extraction methods pertinent to extraction of botanicals, and demonstrates the importance of the choice of appropriate extraction methods for drug discovery and development from botanical sources. It also covers various mathematical models and chemometric aspects in relation to extraction. This book is replete with several practical examples, suitable diagrams and figures, and is easy to follow.

The book comprises 10 well-written chapters, starting with a basic introduction (Chapter 1) and ending with Chapter 10, which covers aspects of profiling crude extracts for rapid identification of bioactive compounds. Although the book mainly focuses on various extraction technologies and methods, it also incorporates chapters on various related matters—for example, Chapter 6 deals with identification strategies of phytochemicals. The profundity of this book relies on the fact that the authors have perfectly utilized their own experience and knowledge in this area of research, and also captured contemporary relevant literature. This book is certainly a valuable addition to currently available classical well-known books in this area, e.g., Phytochemical Methods (By J.B., Harborne), and Natural Products Isolation (Eds: Sarker, S.D. and Nahar, L.).

It is my pleasure to provide this foreword and to recommend this valuable book to all, experienced or novice, who have been involved in research with botanicals.

Professor Satyajit D. Sarker
Editor-in-Chief, Phytochemical Analysis
Director, School of Pharmacy and Biomolecular Sciences
Liverpool John Moores University
Liverpool L3 3AF
United Kingdom

Foreword by Verpoorte

"What you see is what you extract"

Recently we wrote an editorial for a special issue of Phytochemical Analysis on metabolomics with the above mentioned title (Choi and Verpoorte, 2014). In fact the extraction is the most crucial step in any process concerning analysis and isolation of natural products as well as in the use of medicinal plants as such. This book deals with all the aspects involved from harvesting to possible final analyses of the extracts in quite a comprehensive way. In biochemistry and molecular biology, extraction plays a major role for which highly standardized methods are used and in case of the latter field even standard kits are used for DNA and RNA extraction. Genomics, transcriptomics and proteomics are built upon such a solid basis. Unfortunately such an extraction platform does not exist for small molecules, because in contrast with the mentioned omics, where the targeted macromolecules have similar physical properties, the small molecules cover a very wide range of chemo-diversity and thus of physico-chemical properties. As a result there is no single extraction method that can be applied to extract all small molecules from a biological sample. This is a major factor hampering the development of public databases for metabolomics. What solvent to choose is a major concern, as none is able to extract both polar water soluble compounds. Moreover, poorly soluble compounds will be present in an extract at saturation level, but that does not allow a proper quantitation, as the real amount present in the plant could be of higher magnitudes. This book is a rich source of all relevant information to help developing the right protocol for an extraction, from micro to macro scale. But it even goes a step further as it also takes in consideration the preextraction steps. To develop an optimal extraction for any application, one has to deal with many interconnected variables. To be able to deal with such complexity it is necessary to use design of experiment software; the book indeed introduces this approach as an important research tool. Also methods to analyze the extracts for the presence of various phytochemicals are reviewed. So I think the authors did a great job in covering all these aspects; this book will be an indispensable standard work for any person and any laboratory interested in studying natural products, not only in plants, but in any organism!

Prof. Dr Rob Verpoorte
Natural Products Laboratory, IBL, Leiden University
The Netherlands

Preface

Today when different chromatographic methods can provide high resolution of complex mixtures of almost every matrix, from gases to biological macromolecules, and detection limits down to few nanograms or below, the whole advanced analytical process still can be wasted if an unsuitable sample preparation or extraction method has been applied before the sample reaches the chromatographic system. A poorly prepared botanical extract is sufficient to jeopardize even the most powerful chromatographic detection system. The first step in the qualitative and quantitative analysis of medicinal plant constituents is the "extraction" and it is an important step in studies involving the discovery of bioactive compounds from plant materials. Unfortunately, even though most of the research in medicinal plants begins with extraction, but still today not much attention has been paid to this crucial step. Most people in the world perform botanical extraction every day, without bothering about it, when they make a cup of tea, coffee, or other beverage made with hot water or milk. Even scientists working with medicinal plants often carry out extraction as a casual step considering it as only a necessary step toward the more exciting stages of fractionation and isolation and thus the entire focus shifts toward the greed of ending up with a new compound.

A few moments spent thinking about extraction is amply rewarded when one considers what is happening and how this affects the constituents and their amounts in the extract obtained. If extraction is not performed judiciously the subsequent work and results obtained are misleading and this creates a shaky foundation for further studies utilizing them. Moreover, just immediately after extraction another question that comes haunting to the natural product scientists is regarding the postextraction operations which ultimately pave the way for a new bioactive(s). In such a situation dependency on serendipity is heavily relied upon.

The 10 major chapters of this book which is an amalgamation of our research experience will address these issues; trying to explore natural products research from a different prospective so that the dependency on serendipity is reduced and the same can be turned into planned happenstance. The book for the first time tries to link chemometric strategies to the science of optimized and robust botanical extractions. Through this book, we also have tried to show how technology and environment can complement each other by bringing to the readers the principles and applications of green extraction technologies so that with the advancement of science and technology we still can keep the earth a better place to live in for our future generations.

We humbly seize this opportunity to express our sincere thanks to different national and international funding agencies to support us in our journey of natural products research with the aim to find a magical silver bullet against human sufferings. We are also thankful to our respective organizations for their continuous infrastructural and scientific support. Sincere thanks are due to our research team members. Finally we express our deep gratitude to our family members for being supportive in times of stress in compiling this book.

Subhash C. Mandal
Vivekananda Mandal
Anup Kumar Das

Chapter 1

Introduction

Chapter Outline

The subjects "Pharmacognosy" and "Natural Product Chemistry" have developed concomitantly, and both have coevolved as a single distinct discipline and are closely related as well. Both encompass enormous varieties of naturally occurring entities that are synthesized and accumulated by plants or living organisms and mainly deal with chemical structures of these organic entities, their natural distribution, their biosynthesis, rate of turnover and metabolism, and their biological functions or bioactivity.

Pharmacognosy provides the tool to identify, select, and process natural products destined for medicinal and other uses. Although most pharmacognostic and phytochemistry studies focus on plant and phytomolecules derived from them, other sections of organisms are also regarded as pharmacognostically interesting as shown in Figure 1.1.

Crude or an untreated extract from any one of these sources may typically contain known, novel, structurally similar, or diverse chemical(s) with or without some form of biological activity. Generally, bioactive compounds of natural origin from the above-cited sources are popularly known as Secondary metabolites. In order to understand the term secondary metabolites, it is mandatory to understand what is meant by primary and secondary metabolism.

The process of synthesizing essential components in plants such as proteins, carbohydrates, fats, and nucleic acids that are all vital for their sustainability is generally termed primary metabolism, and the essential components thus produced are called primary metabolites. An interesting fact about a plant's defense system against some natural, synthetic factors or unnatural factors (virus attack, radiation exposure) is that plants start producing compounds that are generally not essential for their growth, development, or reproduction but are produced either as a result of the organism adapting to its surrounding environment or as a defense mechanism against predators to help in the sustainability of the organism. Such compounds are called secondary metabolites, and the process through which they are formed is known as secondary metabolism.

Essentials of Botanical Extraction. http://dx.doi.org/10.1016/B978-0-12-802325-9.00001-X

1

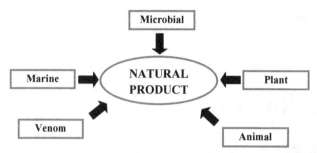

FIGURE 1.1 Various sources of natural products.

The study of Pharmacognosy can be divided into the following fields:

1. *Medical ethnobotany*—this deals with the understanding and study of the traditional uses of plants for medicinal purposes.
2. *Ethnopharmacology*—the study of efficiency and efficacy qualities of traditionally used plants and medicinal substances derived from them.
3. *Phytotherapy*—a part of pharmacognosy focusing on the use of crude or semipure mixtures of extracts for medicinal use. It is sometimes considered as an alternative medicine.
4. *Zoopharmacognosy*—deals with self-medication of nonhuman animals by selecting and using plant parts, soils, and insects to treat and prevent diseases.
5. *Marine pharmacognosy*—study of chemicals derived from marine organisms.

Strategies for research in the area of natural product chemistry and pharmacognosy have evolved quite significantly over the last 20 years. These can be classified into two main categories:

1.1 OLDER STRATEGIES

1. A straightforward chemotaxonomic study.
2. Selection of potent sources simply relying on ethnobotany and ethnopharmacology reports.
3. Simple adoption of phytochemical surveillance and in the process overlooking their bioactivity.
4. Simple isolation and identification of compounds from natural sources followed by *in vitro* and *in vivo* biological action investigation.

1.2 CONTEMPORARY STRATEGIES

1. Modern advances in extraction and separation techniques actually have helped phytochemists to venture and adopt bioassay-guided isolation of secondary metabolites.

2. Strategies to access the metagenome of the botanical source by constructing and screening DNA libraries. Such genome-mining strategies appear to be a promising tool for the discovery of bioactive compounds.
3. Introduction of the concepts of system biology, reverse pharmacology, dereplication, chemogenomics, chemically engineered extract, gene microarray analysis, metabolomic fingerprinting, and chemical fingerprinting actually have shifted the focus more towards bioactivity oriented drug discovery.
4. Introduction of "omic" technologies as high-throughput methods has actually opened the methodological possibilities to investigate extensively into the pharmacological mode of action, synergy effects, multitarget synergy effects, and multitarget property of complex crude extracts.
5. Introduction and advancement of sophisticated hyphenated techniques (high-performance liquid chromatography mass spectrometry (HPLC–MS), HPLC-nuclear magnetic resonance (NMR), HPLC–MS–NMR, HPLC–diode array detector–MS–NMR, etc.) in the field of separation techniques with high-sensitivity detectors have allowed better detection of small molecular compounds present in biological systems.

The advantages and challenges of botanical drug discovery as compared to its synthetic counterpart are summarized below:

1.3 ADVANTAGES

- Botanicals offer an unmatched chemical diversity coupled with immense biological potency.
- Bioactive molecules obtained from nature have evolved to bind to proteins and do possess drug-like properties.
- Bioactive molecules obtained from botanicals can provide various chemical "building blocks" that can be used to synthesize more complex molecules. A popular example has been diosgenin from *Dioscorea floribunda* for the synthesis of oral contraceptives. Similar was the case with camptothecin from *Camptotheca* species, which led the development of novel anticancer molecules such as topotecan and irinotecan.
- Selection and usage of botanical sources are mainly done on the basis of its long-term use by humans and is popularly known as ethnomedicine. This approach is thought to be safer than selecting plant species with no history of human use. Discovery of drugs from *Rauwolfia serpentine*, *Digitalis purpurea*, etc. in the past actually comes under this class of approach of drug discovery.
- Pure bioactives can be administered in a reproducible, accurate dose with apparent therapeutic benefits.
- They can also lead to the improvement of analytical assays for particular compounds or for compound classes. This helps in the screening of plants for potential toxicity and for quality control of therapeutic formulations for human or animal intake.

- Bioactive compounds assist in structural elucidation, which may enable the production of synthetic compounds, incorporation of structure-related modifications, and finally help in the validation of mechanisms of action.

1.4 CHALLENGES

Utilization of the whole plant or their crude preparations for therapeutic or experimental reasons can have several drawbacks including

- Disparity in the amount of the bioactives with different geographic areas, seasons, with different plant parts and morphology, and with different climatic and environmental conditions.
- Co-occurrence of unwanted compounds causing synergistic, antagonistic, or undesirable responses; modifications of the pharmacological activity cannot be disregarded.
- Changes or losses of bioactivity occur due to variability in collection, storage, and preparation of the raw material.
- Lack of an appropriate dereplication strategy has resulted in reoccurrence of species and phytocompounds within many extract libraries. Redundancy from an economical point of view is totally undesirable in natural product drug discovery.
- Production of an adequate amount of the potent bioactive compound is needed for drug discovery and drug profiling, which may require an extensive optimization and scale-up steps.
- The extraction, isolation, and characterization of active compounds from botanicals are an extremely time-consuming and labor-intensive pursuit.
- Lack of systematic exploitation of the ecosystem for the discovery of novel compounds and subsequent commercialization would pressurize the source to a large extent and might lead to detrimental and irreversible environmental effects. Anticancer molecules like topside, docetaxel, and topotecan would continue to depend upon highly vulnerable plant resources for a continuous supply to obtain the starting material as complete chemical synthesis for these phytocompounds is not possible.

Nature recognizes no synthetic barriers such as is represented by "academic disciplines," and thus, it is no surprise to find investigators with quite different academic training studying various aspects of natural product research. It is when such diversely trained investigators come together as a team or, at the very least, collaborate very closely with one another, that most benefit will arise from such studies since the investigators approach the subject from different perspectives, and thus complement each other in their research. The multidisciplinary approach enables us to solve the difficulties of botanical isolation and purification very efficiently. The multidisciplinary approach may also enhance the diversity and consequent value of bioactive natural product research. Successful multidisciplinary collaboration requires that each individual has an understanding and appreciation of the intellectual and technical contributions to the project of the other team members. The sophistication of modern scientific instruments is such that the human aspect of using such equipment is frequently ignored. Although almost any research problem can presently be solved by the application of the most powerful equipment in the market, it requires imagination and resourcefulness to achieve results when such techniques are unavailable or too expensive to employ. It is then essential for the investigators to ask themselves several questions with regard to the most expeditious approach to be implemented, for example,

- Can a single bioassay be implemented or used?
- Is a "value-added" approach absolutely necessary to investigate other potential bioactivities?
- What can be the most valuable method for separation of bioactives for both structural determination and large-scale biological testing? Apparently, it is not possible to develop a separation technique that will produce enough milligram quantity necessary for structural determination followed by developing an entirely different method for preparation of gram quantities for *in vitro* and *in vivo* testing.
- What are the indispensable requirements for an explicit structure determination? Can the presumed structural features be confirmed using sophisticated strategies or analytical techniques?
- Can the data pertaining to separation and molecular structure determination be integrated to give a useful method for detection and analysis?

Scientists working in the field of natural product chemistry must have a basic understanding of both the potential and the limitations of the techniques and disciplines that each can bring to bear on a problem relating to natural product drug discovery. It is the intention of this book to highlight some of the modern sophisticated extraction and separation techniques of plant-derived natural products for a better understanding of the goal. Through this book, we hope to bring these modern techniques, which have not yet received commercial interest, into the limelight so that it can open new exploration methods in search of bioactive phytomolecules.

Several chapters of this book will take the readers on an unexplored scientific and technological ride of herbal extraction methodologies. Even though, some of the extraction methods described in this book are not yet commercialized due to the lack of acceptability toward new things, through this book, we hope to change the thinking of herbal drug researchers so that they can see the new face of herbal drug extraction technology. This book is unique in its own kind, where we have not repeated the orthodox pattern of presentation of extraction schemes of different phytoconstituents. Instead, we have focused fully on the concept delivery of some sophisticated extraction methodologies applicable to herbal drugs and what impact it can have in the quality control and standardization of herbal drugs.

Chapter 2 deals with some of the basic information related to the History and Background on the use of Natural products obtained from plants as therapeutic agents.

Chapter 3 encompasses the various aspects of botanicals as a screening source of new drugs, their historic and present contribution in discovering potent bioactives during 1984–2014.

Chapter 4 shall give the readers a vivid picture about the role of plants in preparing a launch pad for natural product drug development; it shall also try to reveal the reasons for the declining interest in botanical drug discovery and strategies to improve natural products drug discovery from botanicals.

Chapter 5 covers in detail the various aspects pertaining to several strategies, techniques, and methodologies to make botanicals accessible for extraction, which is the prime objective of this book. It covers various approaches in selecting botanicals along with several modern extraction technologies. To add more to this, we have also tried to highlight some basic aspects on extraction theories and various factors affecting extraction.

Chapter 6 highlights several modern extraction techniques in detail, encompassing their working principle and instrumentation with suitable examples.

Chapter 7 highlights a new emerging structured approach for the process development of solvent-based botanical extraction techniques. This chapter will highlight some approaches in the selection of the best optimal conditions that will determine the experimental outcome through a combination of the design of experiments and rigorous modeling technique. Through this chapter, an overview covering various approaches in combination with statistical techniques and chemometrics encompassing botanical background shall be discussed to have an in-depth process understanding for an improved process economy.

Chapter 8 describes the different strategies for the identification of volatile and nonvolatile phytoconstituents.

Chapter 9 describes the different chemical tests applicable to different secondary metabolites.

Chapter 10 covers in detail the strategies of profiling crude extracts for the rapid identification of bioactive compounds. Such strategies will play a major role in identifying lines that will assist in producing the optimal yield of the desired product in the absence of interfering compounds.

Chapter 2

History and Background on the Use of Natural Products Obtained from Plants as Therapeutic Agents

Chapter Outline

2.1 A GENERAL OVERVIEW

About 5000 years ago, when writing did not emerge, knowledge and wisdom continued to pass on simply through word of mouth from generation to generation, or got preserved in the form of epics and poems.

According to some historians, ancient cave paintings on walls may hold clues to the early use of plant-derived drugs. However, it is only written documents that led our ancestors to look out for remedies for treating diseases prevailing at that time, such as those compiled by the inhabitants of Ancient Egypt, Mesopotamia, and particularly Greece.

The notion that cannot be ruled out is that most of the prehistoric societies initially must have adopted a straightforward empirical approach to treat minor ailments of everyday life, selecting healing herbs by a process of trial and error. Persons who were surviving in hot sandy deserts looked out for soothing balms or lotions to apply to their dry skin and eyes, while those suffering from constipation must have sought fiber-containing herbs. Botanicals have often served as the sole mean to treat diseases and injuries for our ancestors either by chewing certain herbs and plants or by simple wrapping leaves with wound-healing properties around the wound or injury. In cases of poisoning or stomach upset our ancestors must have ate or swallowed herbs that irritated the wall of the stomach

Essentials of Botanical Extraction. http://dx.doi.org/10.1016/B978-0-12-802325-9.00002-1

to stimulate vomiting, while for minor cuts or burns rubbing of leaves or barks rich in tannins might have been practised.

All throughout our evolution, the importance of plant-derived bioactive phytocompounds for health and procuring medicinal agents has been enormous. Due to diverse medicinal potentials and biochemical activities of natural products, nearly every civilization and people from different time periods and ethnic groups have accumulated experience and knowledge about their uses.

2.2 DRUG USAGE DURING THE PREHISTORIC PERIOD

Long before the emergence of writing, there was another form of communication that has served to safeguard information and important data about the lifestyle of our prehistoric ancestors. This was rock art and painting, comprising messages on rocks either exposed to the atmosphere or protected inside caves.

Drugs consumed by our Neolithic ancestors are those that exhibited an effect on the mind, such as peyote, mescal bean, alcohol, betel, poppy, coca, cannabis, and tobacco.

In the year 1958, an American anthropologist Thomas Campbell found the American Indian inscriptions of the mescal bean (*Sophora secundiflora*) and drawings on the walls of caves near the convergence point of the Pecos River and the Rio Grande in Texas. Campbell also observed that mescal beans had repeatedly been found in nearby caves, and when radiocarbon dated, the period was predicted to be around 7500 BC to 200 AD. In another cave, he found the remains of a mescaline-containing cactus (*Lophophora williamsii*), and Mexican buckeye (*Ungnadia speciosa*), both known to be psychoactive.

Central Saharan Desert cave paintings from the era of 7000–5000 BC depict scenes of harvesting, respecting, dancing, and offerings to mushrooms.

Seeds of the mildly psychoactive betel nut (*Areca catechu*) were found in the Spirit Cave in the northwest of Thailand and are thought to be there between 7000 and 5500 BC. The earliest direct evidence of human consumption of betel nut comes from the Duyong Cave in the Philippines, where a skeleton from 2680 BC, was discovered buried beside shells containing lime where the teeth of the Duyong Cave skeleton were blackened. This is reminiscent of the practice still prevalent in India, where betel nut is wrapped in a betel leaf (*Piper betel*) and lime is then added (to liberate the readily absorbable free base of arecoline) to it followed by chewing.

Historical evidence revealed that the use of plant-derived drugs was mainly for social and recreational use. But conclusive evidence about their use as medicine does not exist. Therefore, it is not surprising that archaeologists have found more evidence of drugs used for social and recreational purposes than for medicinal purpose. In many findings, it has not only been suggested that the original use of drugs was in religious rituals but it has also been claimed that early religious practices arose from the widespread use of psychoactive drugs.

It is not at all surprising that alcoholic beverages were among the earliest of drugs to have been discovered since their effects must have been readily apparent within a short period of ingestion.

The role of alcoholic beverages in religious practice in ancient times is beyond dispute if we consider the Greek Dionysian festivities, the Roman Bacchanalia, and the importance of wine in Judaism and Christianity.

A 4000 BC seal showing two people drinking beer with a straw (must have been used to filter out the mash floating in the beer) was discovered during 1930 excavations of the Tepe Gawra site near the ancient city of Nineveh in northeast Iraq, and consumption of beer is frequently featured in several third millennium BC Sumerian drawings and texts.

Evidence of wine availability and consumption over 7000 years ago has been obtained from several chemical analyses and testing of yellow residues found in six vessels buried under the floor of a mud brick building of the Neolithic Hajji Firuz Tepe village discovered in the Zagros Mountains of western Iran. Several chemical analyses of the yellow residue indicated that it consisted largely of calcium tartrate together with resin from the terebinth tree. The presence of calcium tartrate could only have come from grape juice as there is no other common natural source. In another excavation at Godin Tepe village, earthenware jars from 3500 to 3000 BC were found where again some deposits of calcium tartrate were reported.

Opium obtained from the poppy, *Papaver somniferum* L. is one of the oldest cultivated species known and is based on the finding of seeds of the closely related to *P. somniferum* subsp. *setigerum* in Germany at the Danubian settlements around 4400–4000 BC. Farming settlements in and around northern France and Switzerland were also known to contain greater quantities of poppy seeds dating back to 3700–3625 BC.

2.3 DEVELOPMENTS AND DRUG USAGE DURING ANCIENT TIMES/PRE-HELLENIC CIVILIZATIONS

- Mesopotamia

The earliest record of botanical use is generally found on clay tablets in cuneiform from Mesopotamia ("2600 BC") where it has been reported that oils from *Cupressus semperviens* and commiphora species were used for the treatment of cough, cold, and inflammatory-related ailments. A major source of information about drugs used in Mesopotamia is provided by the 660 tablets of clay discovered in the year 1920 from the library of a palace of the last king of Assyria in Nineveh. The tablets were the copies of much older texts, with some of the drugs possessing Sumerian names from the third millennium BC.

From the translated work, we learn that there were two types of healers prevalent in Mesopotamia, namely, sorcerers and physicians. The former believed that disease was caused by demons and used magic and incantations to banish them, while the latter followed a therapeutic tradition by using drugs of animal, vegetable, and mineral sources using a more practical approach. From all this findings, it is quite apparent that an independent therapeutic tradition had developed in Ancient Mesopotamia and appears likely that it passed on to the Egyptians.

- Egypt

The ancient Egyptian's *Ebyrus papyrus* dating around "3000 BC," contains about 800–900 complex prescriptions with almost >700 natural herbs in which notably are pomegranate, castor oil plant, aloe, senna, garlic, onion, fig, willow, juniper, common centaury, etc., as internal medicine for eye-, skin-, and gynecology-related complications. Spices such as cassia, caraway, coriander, cumin, fennel, gentian, juniper, peppermint, and thyme were also being used by the ancient Egyptians. Many of the prescriptions that were excavated from Nineveh were meant for skin and eye complaints along with remedies for gastrointestinal diseases. The *Ebyrus Papyrus* has a prescription for *Cannabis sativa* (marijuana) that was meant to be applied topically for inflammation.

Unlike in Mesopotamia, the treatment of the sick in Egypt was in the hands of the physicians who were highly regarded priests. *Imhotep*, the oldest physician known to historians, lived during the Third Dynasty of the Old Kingdom around 2650 BC. At that time, the use of magic was closely entangled with drug use whereby incantations were often being uttered prior to drug administration to make the drug action more effective. The very conception of the influence of psychological elements on the response of a therapy paved the way for the modern knowledge of the "Placebo" effect.

- India

Ayurveda, the "ancient science of life and healthy living," is the holistic alternative science from India, and is >5000 years old. It is extensively practiced in the Indian subcontinent and is also one of the official systems of medicine in India. Its concepts and approaches are considered to have been perfected between 2500 and 500 BC. It is believed to be the oldest healing science in existence, forming the foundation of all others apart from being the comprehensive medical system available. Buddhism, Taoism, Tibetan, and other cultural medicines have many similar parallels to Ayurveda. Historians have often argued about the origin of the Ayurveda whereby some believe that the gathering of healers from around the world in India with their medical knowledge paved the way for Ayurveda where the complete knowledge was preserved in writings along with the more spiritual insights of ethics, virtue, and self-realization. Others argue that Ayurveda was passed down from God to humans through angels (Figure 2.1).

Herbs and mineral drugs used in Ayurveda were later described by ancient Indian herbalists Charaka and Sushruta. It was designed to promote good health and longevity rather than to fight disease and was practiced by physicians and

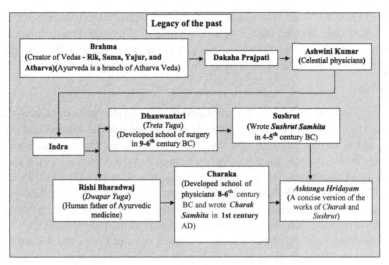

FIGURE 2.1   A journey into the history of Ayurveda.

surgeons called vaidyas. Till "700 BC," this ancient science was vocally discussed between sages and physicians. Thereafter, two different textbooks were compiled: one by "Charaka," which is called *Charaka Samhita* and the other by "Sushruta" popularly known as *Sushruta Samhita*. The former deals with the etiology, symptomatology, pathology, prognosis, and medical management of disease; *Sushruta Samhita* deals with various surgical instruments and procedures. Ayurveda during the course of ancient times developed as a sound scientific system, and this is evident as it is divided into eight major disciplines known as *Ashtanga* Ayurveda. There are eight branches (Ashtanga) of Ayurveda and are as follows:

1. Internal medicine—Kayachikitsa Tantra
2. Surgery—Shalya Tantra
3. Ears, eyes, nose, and throat—Shalakya Tantra
4. Pediatrics—Kaumarabhritya Tantra
5. Toxicology—Agada Tantra
6. Aphrodisiacs/Reproductive Medicine—Bajikarana Tantra
7. Health and Longevity—Rasayana Tantra
8. Spiritual Healing/Psychiatry—Bhuta Vidya

   Herbs such as Amalaki (Indian gooseberry), Amlavetasa (Rhubarb), Arjuna, Ashwagandha, Brahmi (gotu kola), Devadaru (Himalayan Cedar), Gudmar, Guggul, Haridra (turmeric), Ispaghula, Haritaki, Jatamanshi, Kumari (Aloe Vera), Maricha (Black Pepper), Neem, Nirgundi, Pippali, Punarnava, Garlic, Sarpagandha, Sarasparilla, Shatavari, Shankhpushpi, Sesame, Shwetamusali (White Musali), Tulsi, Vasaka, licorice, etc. are very popular and well-known herbs in Ayurveda. Apart from few mineral drugs, several herbal mixtures like Triphala churna (mixture of Amalaki—*Emblica officinalis*, Haritaki— *Terminalia chebula*, and Vibhitaki—*Terminalia bellirica*), Trikatu churna (mixture of *Piper longum, Piper nigrum,* and *Zingiber officinale*), Chyavan Prash, Sitopaladi churna (Tabasheer, *Piper longum, Elettaria cardamomum,* and *Cinnamomum zeylanicum*), Lavan Bhaskar, Yogaraj Guggul are very popular as well.

● China

The Traditional Chinese system of medicine (TCM) has been in existence for at least thousands of years. It includes Han medicine and the theories and practices of the various national minorities, such as the Miao, Dai, Mongolian, and Tibetan nationalities, each with their own unique systems as that of Han medicine. TCM flourished during the era of The Yellow Emperor (2698–2598 BC). The Chinese Materia Medica (1100 BC.) containing 52 prescriptions, Shennong Herbal (~100 BC), containing 365 drugs and the Tang Herbal (659 AD) containing 850 drugs are documented records of the uses of natural products.

Shennong popularly known as the legendary ruler of China and culture hero is one of the Three Sovereigns (also known as "Three Emperors") lived some 5000 years ago. He had even tested the medicinal properties of many herbs on himself even by eating hundreds of plants and even consuming 70 poisons in one day. Emperor Shennong, the founder of Chinese herbal medicine, compiled the "Shennong pen Ts'ao Ching or Great Herbal" around 2700 BC.

Pen Ts'ao, the first compilation of herbal medicines, is connected with his name (Unschuld 1973, 1986).

It is considered to be the earliest Chinese pharmacopoeia, and includes 365 medicines derived from minerals, plants, and animals.

● Greece and Rome

Hippocrates (ca "460–377 BC") had collected >400 natural herbs and their phytocompounds whose importance have been documented in his famous work "*Corpus hippocraticum*." Hippocrates is one of the most renowned figures in the history of medicine, often being called the "Father of Medicine." For the first time in Greeco-Roman history, an account of the four humors—blood, black bile, yellow bile, and phlegm—appeared in the "*C. hippocraticum*."

Theophrastus (371–287 BC) compiled two books, namely, "*De Causis Plantarium*"—Plant Etiology and "*De Historia Plantarium*"—Plant History. Through his astute approach and observation, he generated a classification of >500 medicinal plants known at that time. Apart from these, he referred to cinnamon, false hellebore, mint, iris rhizome, pomegranate, cardamom, fragrant hellebore, monkshood, etc. While describing the toxic actions of plants, Theophrast emphasized the important features for humans to become familiar with them by a steady increase of doses.

Celsus ("25 BC–50 AD") was the first among the Greeco-Roman medical authors, who through his work "*De re medica*" quoted approximately 250 medicinal plants such as aloe, henbane, flax, poppy, pepper, cinnamon, the star gentian, cardamom, and false hellebore.

Sometime between "50 and 80 AD," a Greek physician known as Pedanius Dioscorides, who is generally known as "the father of Pharmacognosy," was a military physician and pharmacognosist of Nero's military, indeed studied medicinal plants wherever he went with the Roman Army. Around approximately 77 AD, he compiled "*De Materia Medica*," a collection of >600 plants, 35 animal products, and 90 minerals. This classical work of ancient history got translated many times, offered plenty of information on the medicinal plants constituting the basic materia medica until the late Middle Ages and the Renaissance. Of the total of 944 drugs described in it, 657 were from plant origin, with descriptions of phenotypes, area of occurrence, collection practices, formulation of medicated preparations, and their therapeutic values apart from discussing its humoral qualities and its characteristics so that adulteration could be recognized. He took great care to include not only as much information as he could about the medicinal uses of the herbs in humans and animals but also reported the side effects. The most cherished domestic plants during his time were as follows: willow (from the genus Salix), chamomile or Camomile (as an antiinflammatory agent), garlic, sea onion (as diuretics), Mentha (to relieve headache and stomach ache), oak bark (gynecological purposes), and parsley to name a few. There were also about 90 mineral drugs, including calamine, mercury, and sulfur. With the gradual passage of time, De Materia Medica strongly influenced Arab physicians after the downfall of Rome. It was among the first books to be printed and has continued to be widely accepted authority until the arrival of the eighteenth century.

Gaius Plinius Secundus ("23–79 AD"), a contemporary of Dioscorides, was a Roman author, naturalist, and natural philosopher, as well as naval and army commander of Roman emperor Vespasian. He had traveled throughout Germany and Spain and wrote about approximately 1000 medicinal plants in his book "*Historia naturalis*." Pliny's and Dioscorides' work incorporated all knowledge of medicinal plants at the time.

Galen ("129–199 AD") was the most influential of all the Greek physicians who had recorded >540 natural product-derived medicines and practiced botanical medicines to demonstrate that herbal extracts not only contain beneficial components but also some harmful ingredients. He had opined that imbalance of the four humors is the cause of all diseases, but went much further by asserting that it is possible to correct humoral imbalances, not only by supplementing deficient qualities with herbs possessing similar properties but also by administering other herbs that had opposing qualities (opposites cure opposites). His approaches toward contemporary therapy at that time also involved polypharmacy whereby numerous plants were being compounded in a complex formulation that became known to the world as "Galencials."

Galen had compiled the first list of drugs having identical action ("parallel drugs"), which are interchangeable—"De succedanus." Galen also had introduced several new plant-derived drugs in therapy that Dioscorides had not described earlier, for instance, *Uvae ursi folium*, used as an uroantiseptic and a mild diuretic even today.

2.4 DRUG DISCOVERY AND DEVELOPMENT DURING THE MIDDLE AGES

The Middle Age, from around 400 to 1500 AD , witnessed the decline of the Roman influence in every aspect worldwide. This was also the time when plague terrorized many parts of Europe. Diseases such as bubonic plague, leprosy, smallpox, tuberculosis, and scabies were out of control and many millions of people succumbed to these diseases.

• The Arab World

During the ninth century, various medical schools of Persia and Arabia, which were generally more advanced than those in medieval Europe at that time acclaimed the Greco-Roman culture and learning, and had translated tens of thousands of their texts into Arabic language for further study and better understanding. In the ninth century, Baghdad became the focus for a revival of humoral medicine and finally as a result of the Muslim territorial expansion that had occurred there got access to the valuable classical texts stored in the monasteries of Egypt, Syria, and Iraq.

Baghdad with a library of 400,000 manuscripts became a great center of learning and scholarship where physicians could study the ancient classics.

The most illustrious of them was Abu al-Qasim al-Zahrawi's (936–1013) al-Tasrif li-man 'ajaza 'an al-ta'lif. This masterpiece was written when Arabian medicine was approaching its zenith. The second part of the 28th volume of this masterpiece is considered to be a treasure of pharmacognosy as it is concerned with the correct manner of handling plant materials for medicinal application, with an emphasis on the various preextraction techniques such as drying and storage. Examples include acacia, aloes, cardamom, colocynth, fleawort, fumitory, galbanum, lily, liquorice root, lycium, mandrake, opium, scammony, sandalwood, spurge, squill, and wormwood. The natural habitat of each herb is described, together with a description of the plant and how to select the appropriate component and at what season. The expression of juices and collection of gums is covered, as is the filtering of decoctions and other liquid preparations. There is information about preparing oils, vinegar, aromatic waters, amber, coral, turpentine, and so forth. The Arabs introduced numerous new plants in their therapy, mostly brought from India, a country they used to have trade relations; on the other hand, the majority of the plants were with real medicinal value. The Arabs also used pepper, aloe, ginger, belladonna, coffee, henbane, saffron, curcuma, cinnamon, rheum, strychnos, senna, etc.

Owing to their trading background, the Arab voyagers had access to plant materials, herbals, medical texts of the Far East such as China and India.

• Role of Churches

At the beginning of the Middle Ages, the skills of healing, cultivation, and harvesting of medicinal plants and preparation of herbal medicines moved to monasteries. During the early Middle Ages, the Benedictine monasteries were the primary source of medical knowledge in Europe and England. Rather than creating a whole array of new information, most of these monastic scholars focused on translating and hand copying ancient Greco-Roman and Arabic works. The monasteries thus became local centers of medical knowledge, and their herbal gardens provided the raw materials for the simple treatment of common diseases and ailments. At the same time, folk medicine in homes and villages continued incessantly, and remedies were often prescribed along with spells, enchantments, divination, and advice of the so-called "wise women" and "wise men." Therapy was based on medicinal plants, which the physicians-monks quite often grew within the monasteries, and a few among them were sage, seed, savory, anise, mint, Greek, and tansy.

Middle Age European physicians consulted the works of "De Re Medica" compiled by John Mesue (850AD), "Canon Medicinae" by Avicenna (980–1037), and "Liber Magnae Collectionis Simplicum Alimentorum Et Medicamentorum" compiled Ibn Baitar (1197–1248), where >1000 medicinal plants are described.

2.5 DEVELOPMENTS AND DRUG USAGE DURING THE LAST PHASES OF THE MIDDLE AGES

The Renaissance period (roughly from the fourteenth to the late seventeenth centuries) laid the foundation for scientific thoughts in medicinal preparations and medical treatments. There were many advances made in anatomy, physiology, surgery, and medical treatments, including hygiene, public health care, and sanitation.

The great journeys undertaken by Marco Polo (1254–1324) in tropical Asia, China, and Persia, along with the discovery of America (1492), and Vasco De Gama's journeys to discover India (1498), resulted in many medicinal plants being brought on to European soil. Conservation through botanical gardens all over Europe for cultivation of those domestic medicinal plants and of the ones imported from the old and the new world became quite common. With the discovery of America later on, the compilation of the materia medica was enriched with the inclusion of a large number of new medicinal plants and notable among them are *Cinchona, Ipecacuanha, Cacao, Ratanhia, Lobelia, Jalapa, Podophylum, Senega, Vanilla, Mate*, tobacco, red pepper, etc.

In the seventeenth century, quinine bark *Cinchona succirubra* was introduced to the European people. Quinine bark rapidly became famous in France, England, and Germany despite the prevalence of several controversies about its use among distinguished physicians.

Linnaeus during the eighteenth century in his work *Species Plantarium* (1753) provided a brief description and classification of the species described until then. The species were illustrated and named without taking into consideration whether some of them had previously been described somewhere. For naming purpose, a well-coordinated polynomial system was used where the first word denoted the genus while the remaining phrase explained other characteristics of the plant. Linnaeus later altered the naming system into a binominal one. The name of each species consisted of the genus name, with an initial capital letter, and the species name, with an initial small letter.

The early nineteenth century has seen a lot of changes with respect to knowledge and use of medicinal plants. The discovery followed by isolation of alkaloids from poppy in 1806, ipecacuanha in 1817, strychnos in 1817, quinine in 1820, and pomegranate in 1878 to name a few.

- Digitalis: Extracted from a plant called foxglove; digitalis stimulates the cardiac muscles and was used to treat cardiac conditions. In the late 1700s, William Withering introduced digitalis, an extract from the plant foxglove, for treatment of cardiac problems.
- Quinine: Derived from the bark of the Cinchona tree, quinine was used to treat malaria.
- Ipecacuanha: Extracted from the bark or root of the Cephaelis plant, ipecacuanha was used to treat dysentery.
- Aspirin: Extracted from the bark of willow tree, aspirin was used for the treatment of fever.

More systematic research was being performed to discover new drugs from the early 1900s.

In the late half of the nineteenth and early twentieth centuries, there was a great risk of omission of medicinal plants from therapy to treat ailments. Many authors were of the opinion that drugs obtained from medicinal plants had many disadvantages due to the destructive action of enzymes during their drying and storage. During the nineteenth century, alkaloids and glycosides isolated in pure form were increasingly replacing the drugs from which they had been isolated. However, it was soon ascertained that although the action of pure alkaloids was faster, the action of alkaloid drugs was full and long lasting. In the early twentieth century, stabilization methods for fresh medicinal plant materials were proposed, especially for the thermolabile components. Besides, much emphasis was invested in the study of manufacturing and cultivation condition for medicinal plants.

From the early 1930s, drug discovery concentrated on screening natural products and isolating the active ingredients for treating diseases. The active ingredients are normally the synthetic version of the natural products. These synthetic versions, called new chemical entities, have to go through many iterations and tests to ensure they are safe, potent, and effective.

At present, almost all pharmacopoeias in the world, including the European Pharmacopoeia, The United States Pharmacopoeia, and British Pharmacopoeia, do not include plant drugs. Several countries like India, China, United Kingdom, Russia, Japan, and Germany do have their separate herbal pharmacopoeias.

In each and every phase of human civilization, the healing properties of certain medicinal plants were identified, noted, and conveyed to their successive generations. The benefits of one society were passed on to another whereby upgradation of old properties, theories, and beliefs took place. The continuous and perpetual interest of people in medicinal plants has brought about today's modern and sophisticated fashion of their processing and usage for the benefit of humankind.

FURTHER READING

Smith, C.G., Donnell, O.J.T. (Eds.), 2006. The Process of New Drug Discovery and Development. second ed. Informa Healthcare.

Wiart, C., 2006. Etnopharmacology of Medicinal Plants. Humana Press, New Jersey. 1–50.

Chapter 3

Botanicals as a Screening Source of New Drugs: Past Success Stories and Present-Day Concerns

Chapter Outline

3.1 HISTORIC ROLE OF BOTANICALS

Over the centuries, human civilizations have banked on plants and plant parts (wood, leaves, fibers, fruits, tubers) for their basic needs such as food, clothing, and shelter. Plants have also been used for some supplementary purposes, namely, as dart and arrow poisons for hunting, psychoactives (hallucinogens, and narcotics) for performing rituals and ceremonies, poison for murdering, as stimulants for stamina and vigor, as hunger suppression, and lastly noteworthy to mention is for medicines. Phytochemicals that were used as medicines are mainly secondary metabolites.

- For Hunting
  Arrow poisons were used to poison the heads of arrows or darts for the purpose of hunting and are commonly practiced now in the areas of South America, Africa, and Asia. In southeast Asia, such things are practiced in India (Assam), Burma, and Malaysia.
  Sources for arrow poisons are mainly members of the Strychnos and Strophanthus genera. The poison ingredient (strychnine or strophanthin) attacks the central nervous system causing paralysis, convulsion, and cardiac arrest. These agents actually block the action of acetylcholine (a neurotransmitter, which controls muscle contraction) and prevent it from attaching with the nicotinic acetylcholine receptor thus blocking muscle contraction and the prey eventually dies due to paralysis. Most compounds responsible for the potency of arrow and dart poison belong to three plant chemical groups,

Essentials of Botanical Extraction. http://dx.doi.org/10.1016/B978-0-12-802325-9.00003-3
19

namely, the alkaloids (e.g., curare from *Chondrodendron tomentosum*, strychnine from Strychnos species), cardiac glycosides (ouabain from Strophanthus species), and saponins (sapotoxin from soaproot). In 1939, the active ingredient of curare was isolated. In 1943, it was introduced effectively into anesthesiology. Curare provided satisfactory muscle relaxation without the depressant effect, which is induced by deep anesthesia induced by ether or chloroform. The active ingredient of curare, D-tubocurarine, led to the discovery and development of various new skeletal muscle relaxants and one among them is intocostrin used in surgery ever since (Box 3.1).

Box 3.1 Some Interesting Points about Curare

- For many centuries, the exact content of curare remained a mystery to Western observers, not until 1800 when Alexander Von Humboldt witness and document the preparation of curare by the Indians residing near Orinco River.
- In 1814, an explorer named Charles Waterton upon injecting a donkey with curare found that within 10 min, the donkey appeared dead. Upon cutting a small hole in her throat and inserting a pair of bellows to blow air for inflating the lungs, the donkey held her head up and looked around. Waterton then continued artificial respiration for 2 h until the effects of curare had worn off. Later on after scientifically studying curare, it was found to block the transmission of nerve impulses to muscle, including the diaphragm muscle, which controls breathing and if unchecked might cause death.

- As poison for Murdering

  Since ancient times, many plants have been recognized for their poisonous properties, and people have used them to induce homicide. During ancient times, sometime between 3100 and 3000 BC, Menes, probably the first pharaoh of both upper and lower Egypt, thoroughly studied and investigated several poisonous plants and came to know about cyanide content in the leaves and pits of the peach plant (*Prunus* spp.). The Egyptians were also well acquainted with many other poisonous plants as well, including opium (*Papaver somniferum*) and mandrake (*Mandragora officinarum*), the latter belonging to the Solanaceae or nightshade family. Greek philosopher Socrates was sentenced to death by forcing him to drink a concoction of water hemlock (*Cicuta maculata*, belonging to the family Apiaceae, commonly known as cowbane). The use of poisonous plants was also practiced by the ancient Romans whereby they used the leaves and berries of belladonna (*Atropa belladonna*), also known as "deadly nightshade," which is another member of the Solanaceae family.

  During the renaissance period, plant poisons came into even more widespread use, particularly favored were aconite (Aconitum, also known as wolfbane and monkshood), belladonna, hemlock (*C. maculata*), jimson weed (*Datura stramonium*), and strychnine (*Strychnos nux vomica*). In 1589, Giovanni Battista Porta compiled Neopoliani Magioe Naturalis, a book compiling the use of poisons from plants, including aconite, almonds (cyanide), and yew seeds (*Taxus baccata*). During the seventeenth century, poisons were in fashion to kill royal family members with the likes of Queen Elizabeth with opium (*P. somniferum*). Some of the plant and its part well documented for murder are mandrake (*M. officinarum*) hemlock (*C. maculate*), belladonna (*A. belladonna*), aconite (*Aconitum napellus*), and strychnine (*Strychnos nux vomica*). In fact, many compounds isolated from poisonous plants were later developed as therapeutic drugs due to their desirable pharmacological actions.

- As a source of Psychoactive drugs
  - As Hallucinogens
    A hallucinogen is any chemical substance that induces a dreamlike state and produces changes (distortions) in perception, thought, and mood that depart from reality. Peyote, marijuana (Cannabis), and lysergic acid diethylamide are examples of hallucinogens. There is much evidence from recorded history that naturally occurring hallucinogens from plants were used in religious, rituals as well as in medicine, sorcery, and for magical applications. The first physical evidence of hallucinogen use can be confirmed from the archaeological findings of traces of nicotine in a 1300-year-old vessel, which belongs to the Mayan civilization. Gas chromatography mass spectrometry and liquid chromatography mass spectrometry confirmed the presence of nicotine in clay vessels popularly known as "house of tobacco," as indicated by hieroglyphic texts. In the old world, humans found some of the most novel means of administering hallucinogens. In southern Africa, the Bushmen of Botswana used to absorb the active constituents of the plant kwashi (*Puncratium trianthum*) by making an incision on the scalp and rubbing the juice of the onion-like bulb into their open wound. In Siberia, the fly agaric (*Amanita muscaria*), a psychoactive mushroom may be toasted on a fire or made into a decoction with reindeer milk and wild blueberries. During the medieval ages, the witches commonly rubbed their bodies with hallucinogenic ointments. Most hallucinogens are alkaloids, a family of approximately 5000 complex organic molecules, which also accounts for the biological activity of most toxic and medicinal plants. These active compounds may be found in various concentration ranges in different parts of the plant—leaves, roots, seeds, flowers, and barks, which all may be absorbed by the human body in a number of ways, as is evident in the wide variety of folk preparations. Hallucinogens may be snuffed or smoked, swallowed dry or fresh, drunk as infusion or decoctions, can be placed in wounds or can be administered as enemas. Tribes of Mexico discovered that sundried cactus Peyote (*Lophophora williamsii*) can be consumed or eaten whole as they produce spectacular psychoactive effects. Mazatec people who are the indigenous inhabitants of Northern Oaxaca, Mexico, discovered a mushroom flora that were

hallucinogenic. Tribes of Venezuela and adjacent Brazil have provided several exceedingly important and chemically fascinating hallucinogenic preparations, notably among them are the intoxicating yopo (*Anadenanthera peregrina*) and ebene (*Virola calophylla, Virola calophylloidea, Virola theiodora*). Several well-recognized plants that contain hallucinogens or psychoactive substances are *Cannabis sativa* L. (Δ9 hydrocannabinol), *L. williamsii* (Mescaline), *P. somniferum* (morphine), and *Datura* species (scopolamine). Several of these plants are still used as therapeutic agents due to the presence of some desired pharmacological activities, and some of the constituents have been developed into modern medicine, either in natural form or as leads.

- As Stimulants and narcotics

   A stimulant is a substance that affects the central nervous system and enhances brain activity in humans by increasing alertness, energy level, and physical activity. Cocaine, caffeine, ephedrine, nicotine, ginseng, and ginkgo are well-known, plant-derived stimulants. A narcotic is a drug that induces central nervous system depression, resulting in numbness, lethargy, and sleep. This includes opiates, alcoholic beverages, and kava. By this definition, nicotine and the stimulant cocaine would also be included as a narcotic. The use of plants as stimulants and narcotics was widely spread all throughout the world in connection with the practice of necromancy and in religious rituals and ceremonies. Anthropological evidence and personal accounts indicate that humans have been using psychoactive substances for many millennia. This is a list of just a few of the most ancient uses of psychoactive substances:

   - **7000 BC: Betel seeds** (*Areca catechu*) were chewed for their stimulant effects; their evidence was found from the several archaeological sites in Asia.
   - According to Raghavan and Baruah 1958, archaeological evidence is present to claim that betel nuts have been used as a narcotic for thousands of years in southeast Asia. The "Spirit Cave" site in Thailand has yielded paleobotanical remains of *A. catechu*, Piper betel (betel leaves), and edible lime, thus providing a circumstantial evidence for the practice of betel chewing in prehistoric times. These remains are between 7500 and 9000 years old. If the dating is correct, this would make betel one of the earliest known psychoactive substances to be used by humans.
   - **6000 BC:** The "tobacco" plant is believed to have been cultivated and harvested as a plant since around 6000 BC. It grows natively in North and South America, and the natives of the Americas had been using tobacco for centuries before the Europeans did. According to archeological evidence, tobacco was the first narcotic used in South America. In the year "1492," when Columbus arrived on the island of Cuba, he and his men observed that the natives were "drinking smoke," by placing

burning rolled up leaves into their nostrils and inhaling them as snuff. The plant was then brought back to Spain, and soon the Europeans believed that tobacco could cure anything. Its use was soon introduced into the Spanish Peninsula, and about 1560, the French ambassador at Lisbon, Jean Nicot, sent some of the fragrant herb to France, where it was named in honor of him "Nicotina."

- **5500 BC:** The first texts about "alcohol" were discovered near 5500 BC. Intentional wine making is believed to have begun in Transcaucasus, east of Turkey or in northwestern Iran. Around the same time, the Chinese were also engaged in making wine with rice and local plant food. Wine making is believed to have been refined through trial and error. One of the biggest challenges to overcome was to manipulate the yeast that turns grape juice into wine and the bacteria that transforms it into vinegar. Earlier wines were mixed with pungent tree resins, probably to preserve the wine in the absence of corks or stoppers. A resin from the ternith tree, a kind of pistachio, was found in wine and dated back to 5500 BC.

- **5000 BC:** Archaeological evidence indicates the use of psilocybin-containing "mushrooms" in ancient times. Hallucinogenic mushrooms have been a part of human culture as far back as the earliest recorded history. Ancient paintings of mushrooms dating to 5000 BC have been found in caves on the Tassili plateau of Northern Algeria. Peoples from the central and southern American cultures had also built temples to mushroom gods carved with "mushroom stones." Mushroom stones and motifs have also been found in Mayan temple ruins in Guatemala.

- **4200 BC:** Archaeological studies revealed the presence of "Poppies" during the prehistoric settlements in Southern Germany, Central Europe, Switzerland, and Southern England during 4200 BC. Other traces of opium poppy were subsequently also identified in various other locations across Europe during the Iron Age, including in England and Poland. During the Iron Age, it was also present in more northerly regions of the world such as the British Isles and Poland. During prehistoric times, the seeds of the opium poppy may well have been used in baking, and their oil been pressed into use for cooking or lighting up lamps.

- **3500 to 3000 BC:** During the period 3500–3000 BC, the people of Bronze Age cultures of the eastern Mediterranean area were consuming wine from metal vessels. During 1930s, an excavation of the Tepe Gawra site near the ancient city of Nineveh in northeast Iraq, consumption of beer was frequently featured in several third millennium BC Sumerian drawings and texts. Evidence of wine availability and consumption has been obtained from several chemical analysis and testing of yellow residues found in six vessels buried under the floor of a mud brick building of the Neolithic Hajji Firuz Tepe village discovered in the Zagros Mountains of western Iran. Several chemical analyses of the yellow residue indicated that it consisted largely of calcium tartrate together with resin from the terebinth tree. The presence of calcium tartrate could only have come from grape juice as there is no other common natural source. In another excavation at Godin Tepe village, earthenware jars from 3500 to 3000 BC were found where again some deposits of calcium tartrate were reported.

- **3000 BC:** The Native American adoration of the "peyote cactus" is centuries old. During an excavation in 2005, researchers used radiocarbon dating and alkaloid analysis to study two specimens of peyote found in archaeological digs from Shumla Cave in Texas. The results indicated that the specimens belong to an era between 3780 and 3660 BC. The main psychoactive compound found in peyote is mescaline. This naturally occurring alkaloid acts as a partial agonist in order to activate the serotonin 5-HT2A receptor. Researchers have shown that activation of this particular serotonin receptor creates a state of psychedelia.
- **3000–2700 BC:** The Chinese Emperor Fu His (c. "2900 BC"), whom the Chinese acknowledge for bringing civilization to China, seems to have made reference to "Cannabis." It was a very popular medicine that possessed both yin and yang. The Chinese emperor Shen Nung (c. "2700 BC"; also known as Chen Nung) who is considered to be the Father of Chinese medicine discovered the healing power of marijuana. It was recommended for constipation, malaria, absentmindedness, rheumatic pains, and female disorders. The earliest written reference about medical utility of marijuana in the Chinese Pharmacopeia is thought to be during the period "1500 BC."
- In Egypt, pollens of cannabis were found on the mummy of Ramesses II, who died in "1213 BC." Prescriptions for cannabis in Ancient Egypt were mainly for the treatment for glaucoma, inflammation, cooling the uterus, and for giving enemas.
- Cannabis in India has been suggested to quicken the mind, induce sleep, lower fevers, stimulate appetite, cure dysentery, relieve headaches, improve digestion, and cure venereal disease. Around "1000 BC" in India, a cannabis drink (Bhang) generally mixed with milk, was used as an anesthetic and antiphlegmatic.
- In ancient Greece (200 BC), cannabis was used as a remedy for earache, edema, and inflammation.
- **2500–2000 BC:** Traces of "coca" have been found in mummies dating back 3000 years in the south-central Andes. The use of coca in ancient Andean life was an important symbol of cultural identity during any type of ceremonies and was used for its medicinal values—to get relief from fatigue, hunger, beneficial in gastrointestinal disorders, arthritis, sores, asthma, and fractures.

Medicinal plants have become the basis of a sophisticated traditional system of medicinal practices that have been used for centuries by people in India (Ayurveda), China (Traditional Chinese medicine), Arab (Tebbe sonnati), Graeco-Arabic (Unani-Tibb or Unani medicine), Australian and Southeast Asian medicine, Africa (Traditional African medicine), and in many other countries.

- As Folklore medicine
  - Indian Folklore medicine
    - *Acacia catechu* Willd. "khair": Santhal tribes make a paste of root and apply it on joints for seven days for rheumatism. Kols of Utter Pradesh use its leaves for blood dysentery.
    - *Bauhinia purpurea* Linn: Santhals, Bhumij, Birhors, and Kherias of West Bengal apply the paste of its bark on sores of small pox. Nagas of Nagaland use its bark for curing cancerous growth in the stomach (locally known as "Chapo").
    - *Cassia auriculata* Linn.: Inhabitants of Maharashtra use its root extract for rheumatism pain. The roots are mixed with *Maytenus emarginatus* roots. Tribals of Eastern Rajasthan use the extract of its seeds for asthma.
    - Camel thorn—The Indian traditional system of medicine claimed that this plant aids in the treatment of dermatosis, fever, anorexia, leprosy, constipation, obesity, and epistaxis due to the presence of a sweet, gummy material mainly melezitose, sucrose, and invert sugar secreted from the stems and leaves during hot days. The Konkani people of India smoked the plant for the treatment of asthma, while the Romans used the plant for nasal polyps.
    - Dhatura (*Datura metal*): Inhabitants of Rajasthan and West Bengal use Dhatura for leprosy.
    - Ashwatha (*Ficus religiosa*): Rabha tribes of West Bengal use its bark for hematuria. Inhabitants of Jammu & Kashmir consume the extract of its bark to relieve whooping cough.
    - Punarnava (*Boerrhavia diffusa*): Inhabitants of Garhwal Himalaya use its roots for piles. The Bhils of Jhabua district in Madhya Pradesh use its roots paste to cure blood dysentery.

  - Native American Folklore medicine
    - Plants genus Salvia—Indian tribes of southern California used to "cook" male newborn babies in hot Salvia ashes as it was believed that these babies would become the strongest and healthiest members of their respective tribes and are claimed to have been immune from all respiratory ailments for their entire life.
    - Plant *Ligusticum scoticum*: Generally abundant in Northern Europe and North America. It was eaten raw in the morning empty stomach to protect a person from daily infection; the root was a cure for flatulence, an aphrodisiac and was used as a sedative in the Faeroe Islands, which

is now an island group and archipelago under the sovereignty of the Kingdom of Denmark.

- Gentian—The Catawba Indians steeped the roots in hot water and applied the hot fluid on aching backs.
- Yellow-Spined Thistle—The Kiowa Indians boiled yellow-spined thistle blossoms and applied the resulting liquid on burns and skin sores.
- Partridgeberry (*Mitchella repens*)—The tribes of Cherokee (Federally recognized tribes from Oklahoma, Texas, California, and Oregon) used a tea of the boiled leaves. Frequent doses of the tea were taken in the few weeks preceding the expected date of delivery to speed up childbirth.
- Roots of Sarsaparilla—The Penobscots (indigenous people of the Northeastern Woodlands, located in Maine) pulverized dried sarsaparilla roots and combined them with sweet flag roots in warm water and used the dark liquid as a cough remedy.
- Roots of Wild Carrot—The Mohegans (federally recognized tribe living on a reservation in the eastern upper Thames River valley of south-central Connecticut) steeped the blossoms of this wild species in warm water when they were in full bloom and drank this to cure diabetes.
- Blossoms of hops (*Humufus lupulus*)—The Mohegan Indians reportedly used an infusion of blossoms of hops as a sedative to relieve nervous tension, whereas the Delaware of Oklahoma used a decoction of leaves of the same plant as a stimulant.

- African Folklore medicine
  African traditional medicine is the oldest and perhaps the most diverse of all medicine systems known. The great biodiversity in the tropical forests, savannahs, and unique environments of sub-Saharan Africa has provided indigenous cultures with a diverse range of plants and as a consequence a wealth of traditional knowledge about their use for medicinal purposes.
  An African traditional medicine follows a holistic approach taking account of both the body and the mind. The healers after diagnosing treat the psychological basis of an illness before prescribing medicines to treat the symptoms. Some of the famous medicinal plants used in Africa for treating ailments are as follows:
  - *Cryptolepis sanguinolenta*: In local traditional medicine, the macerated roots are used as antipyrectic (colic) and hypotensive agents and as a tonic for gastrointestinal problems and against rheumatism. In Ghana, the bark of the root is used in folk medicine to increase virility. Root decoction has been used by traditional healers for the treatment of several fevers (malaria, infections of stomach) and the leaves as an antimalarial and for the cicatrizing of wounds.
  - *Cinnamomum camphora*: In African traditional medicine, camphor is used to take care of coughs, malarial fever (leaf extract), fever due to flu

(bark tea), malaria (leaf infusion is inhaled), and as an antiseptic, stimulant, counterirritant, analeptic, and carminative.

- *Mondia whitei*: In all countries and across all tribes in Africa, mondia finds itself as an aphrodisiac. In Cameroon, the fresh root bark is used to increase the libido, and in Ghana, to increase sperm production.
- *Voacanga Africana*: In Ivory Coast, this plant is used against leprosy, diarrhea, generalized edema, convulsions in children, madness (Tan et al., 2000), as a diuretic, and infant tonic (Iwu, 1993). In southeastern Nigeria, the plant is featured in many healing rituals (Iwu, 1993), including some to induce hallucinations and trances in religious rituals. In Congolese traditional medicine preparations of extracts containing *V. africana* are used as an antiamoebial.

Apart from the above-mentioned medicinal plants, some other famous African medicinal plants include Acacia senegal (Gum Arabic), Agathosma betulina (Buchu), Aloe ferox (Cape Aloes), Aloe vera (North African Origin), Artemisia afra (African wormwood), Aspalanthus linearis (Rooibos tea), Boswellia sacra (Frankincense), Catha edulis (Khat), Commiphora myrrha (Myrrh), Harpagophytum procumbens (Devil's Claw), Hibiscus sabdariffa (Hibiscus, Roselle), Hypoxis hemerocallidea (African potato), Prunus africana (African Cherry), and *Catharanthus roseus* (Rosy Periwinkle).

- Australian and Southeast Asian folklore medicine
  The Aborigines of Australia had a complex healing system, but much of the traditional knowledge in Australia was lost before it could be systematically recorded.
  However, even in the 1960s, there were >124 different plant species that were believed to have therapeutic qualities.
  - Pulverized roots of "Wattle tree" (a tree in the genus "Acacia"): It was used to treat coughs, colds, and laryngitis by soaking them in water to make an aqueous syrup. Aborigines in Tasmania used the rubbery leaves of a plant known as "pig face" to induce vomiting.
  - **Eucalyptus tree:** It was used for gastrointestinal upset and fever. The tree's resin was taken to relieve constipation, and the gum was also powdered and applied to skin sores, with or without an overlay of the inner bark.
  - **Custard apple:** Toothaches were treated by packing wads of inner bark from the custard apple against the decayed tooth. Other concoctions were prepared to treat earaches, dysuria, and several skin diseases.

In contrast, many healing places like Southeast Asian (countries, e.g., Malaysia, Thailand, and Vietnam), New Zealand, Borneo, and the Polynesian Islands remain intact and are being recorded and developed. A strong Chinese influence is being observed in most of these countries. Among the well-known medicinal herbs originating from this region are *Croton tiglium* (Purging croton), *Duboisia hopwoodii* (Pituri), *Eucalyptus globulus* (Bluegum), *Melaleuca alternifolia*

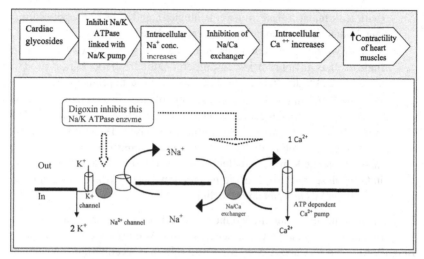

FIGURE 3.1   Mechanism of action of cardiac glycosides.

(Tea tree), *Myristica fragrans* (Nutmeg and Mace), *Piper methysticum* (Kava kava), *Strychnos nux vomica* (Strychnine), *Styrax benzoin* (Benzoin), and *Syzygium aromaticum* (Cloves).

- As a life-saving phytomedicine
  The various discoveries leading to the use of pure drug substances from natural products occurred in the eighteenth and nineteenth centuries. These involved the study of plant preparations known historically from traditional practices to have medicinal properties. Some of the compounds that emerged from the study of ethnobotany and became popular as potent therapeutics for the mankind are as follows:
- Digoxin (a purified cardiac glycoside) from Foxglove (*Digitalis purpurea*), 1785—William Withering was the one who first published his treatment of heart patients with cardiotonic foxglove extract, also called as digitalis. It is useful in arrhythmia and congestive heart failure to increase the force of contraction of the heart muscle. Since the thirteenth century, the herb Foxglove has been used to rinse wounds, and its dried leaves were brewed by Native Americans to treat leg swelling caused by heart problem (Figure 3.1).
- Morphine from poppy (*P. somniferum*), 1804—First isolated by Freidrich Serturner. This discovery enabled an understanding of the opiate receptor subtypes and ultimately an understanding of the endorphin and enkephalin pathways in pain pathophysiology. It has been used for thousands of years to produce euphoria, analgesia, sedation, relief from diarrhea, and cough suppression. It was used medicinally and recreationally from early Greek and Roman times. Opium and laudanum (opium combined with alcohol) were used to treat almost all known diseases (Box 3.2).

Box 3.2 Some Interesting Facts about Opium

- *Frederic Serturner* had tested morphine on himself and three young men and observed that morphine caused cerebral depression and relieved toothache. Gay Lussac named this drug, which was the first discovered alkaloid, after Morpheus (the son of Somnus), morphine.
- "Opium" is a Greek word meaning "juice," or the exudate from poppy.
- "Opiate" is a drug extracted from the exudate of poppy and mainly a naturally occurring alkaloid.
- "Opioid" is a natural or synthetic drug that binds to opioid receptors to produce agonist effects.
- It was in the 1870s that crude morphine derived from the plant *Papaver somniferum* was boiled in acetic anhydride to yield "diacetylmorphine" (heroin) and found to be readily converted to "codeine" (painkiller).
- Historically, it is documented that the Sumerians and Ancient Greeks used poppy extracts medicinally, while the Arabs described opium to be addictive.

- Aspirin from Salicylic acid in genus *Salix alba* L., 1897—Felix Hoffmann working with Bayer Company in 1897 actually synthesized aspirin from salicylic acid present in willow bark. Probably the most famous and well-known example to date would be the synthesis of the antiinflammatory agent, acetyl salicylic acid (aspirin) derived from the natural product, salicin isolated from the bark of the willow tree *Salix alba* L. This innovation is an early example of a synthetic drug inspired by a natural product that had been isolated from a plant preparation with rich ethnobotanical background. Aspirin (Acetyl salicylic acid) can be prepared by the esterification of the phenolic hydroxyl group of salicylic acid with the acetyl group from acetic anhydride or acetyl chloride. In the gut, aspirin is short lived and rapidly gets deacetylated to deliver salicylic acid (Box 3.3).

Box 3.3 Some Interesting Points about Salicylic Acid

- Herman Kolbe synthesized salicylic acid from coal tar. The method is known as "Kolbe Synthesis" in organic chemistry.
- John Vane, a British scientist and Professor was the first to explain that Acetyl salicylic acid actually works by blocking an enzyme called Prostaglandin, which is responsible for pain during tissue injury. In 1982, Professor Vane, now known as Sir John Vane won the noble prize in medicine for this discovery.

The colossal influence of ethnobotany and natural products on drug development was once again affirmed and consequently various new nonsteroidal antiinflammatory drugs(acetoaminophen, ibuprofen, naproxen etc.) were discovered.

- Quinine from the bark of the cinchona tree—The medicinal property of the cinchona tree was originally discovered by the Quechua, who are the natives of Peru and Bolivia; later, the Jesuits (members of the Christian male religious order of the Roman Catholic Church) were the first to bring cinchona to Europe. The bark of the cinchona tree has been used in unextracted form by Europeans since at least the early seventeenth century. In 1631, it was first used to treat malaria in Rome. Before 1820, cinchona bark was first dried, ground into a fine powder, and was then mixed with a liquid (commonly wine) before drinking.

  It was not until the nineteenth century that humans began to isolate the active principles of medicinal plants, and one particular landmark was the discovery of quinine from Cinchona bark by the French scientists Caventou and Pelletier. Such discoveries led to an interest in plants from the New World and expeditions in the almost impenetrable jungles, and forests were undertaken in the quest for new medicines. In 1820, quinine was extracted from the bark, isolated, and named by Pierre Joseph Pelletier and Joseph Caventou. Purified quinine then replaced the bark as the standard treatment for malaria. Quinine and other cinchona alkaloids including cinchonine, quinidine, and cinchonidine are all effective against malaria (Box 3.4).

Box 3.4 Interesting Facts about Quinine in Drinks

- For hundreds of years, quinine has been looked upon as an antidote for fever.
- A TONIC drink in the form of quinine water for the first time was introduced in the British colonies of India in the seventeenth century.
- The British used to mix quinine with a small amount of gin and lemon juice to make it more suitable for drinking. The original idea of tonic water is thought to have come from this.

- Pilocarpine from *Pilocarpus jaborandi*:

  The discovery of pilocarpine for curing glaucoma and dry mouth syndrome could have come much earlier had we closely followed the folk remedies of the indigenous people of Brazil. Upon holding the leaf up to the light, some translucent droplets on its surface are seen. Each droplet is a gland that secretes an alkaloid-rich oil called pilocarpine, a weapon against the blinding disease glaucoma and has been used as a clinical drug in the treatment of chronic open-angle glaucoma and acute angle-closure glaucoma for >100 years.

  In "1994," an oral formulation of pilocarpine was approved by the Food and Drug Administration to treat dry mouth (xerostomia), which is a side effect of radiation therapy for head and neck cancers. In 1998, the oral preparation was approved for the management of Sjogren's syndrome, an autoimmune disease that damages the salivary and lacrimal glands (Box 3.5).

Box 3.5 Mechanism of Action of Pilocarpine

Pilocarpine assists in the transmission of impulses from autonomous nerve endings to the working muscles. These nerves trigger functions such as the beating of the heart and the focusing of the eye. When applied to the eye of a person suffering from early stages of glaucoma, pilocarpine stimulates the muscle that contracts the pupil to relieve eye pressure. Since the disease blinds by building up pressure until the eye can no longer function, pilocarpine can save eyesight.

3.2 BOTANICALS AS SOURCES OF NEW LEADS DURING 1984–2014

Covering a time span of 30 years (1981–2012) Cragg and Newman (2012) in their extensive review asserted that about 26.28% of the 1130 New Chemical Entities were either natural products or were derived from natural product usually by semisynthetic modification and with another about 21.76% are created around a pharmacophore from natural product. In their review, they classified natural products according to their origin using some major categories and subcategories. They are as follows:

- B—Biological, usually a large peptide or protein either isolated from an organism/cell line or produced biotechnologically in a surrogate host.
- N—Natural product
- NB—Natural product "Botanical" (recently approved).
- ND— Natural product derived and is usually a semisynthetic modification.
- S—Totally synthetic drug, often found by random screening/modification of an existing agent.
- S*—Made by total synthesis, but the pharmacophore is/was from a natural product.
- V—Vaccine
- N/M—Natural product mimic

Since the very beginning of human history, chemical compounds produced by plants have been investigated and utilized by physicians to alleviate diseases. Botanicals have been a prolific source of new drugs and leads since the Vedic period. The World Health Organization postulates that about 80% of the inhabitants from developing countries of the world rely on the traditional system of medicine for their primary health care need, and about 85% of traditional medicine involves the use of plant-derived extracts. This means that about 3.5–4 billion people throughout the world depend on plants as sources of drugs (Figure 3.2).

China and India are the two of the largest users of medicinal plants in the world. Over 5000 plant species are used in Traditional Chinese Medicine, whereas India uses about 7000 species. According to the Export Import Bank, the growth rate of 7% per annum has been seen in the international market for

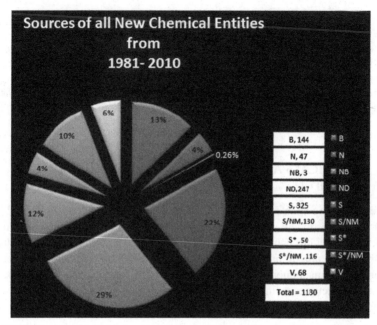

FIGURE 3.2   A graphical representation of the contribution of natural products to drug discovery.

medicinal plant related trade. In the World herbal market, China's share is US$ 6 billion, whereas India's share is merely US$ 1 billion. All the major herbal-based pharmaceutical companies are showing a constant growth of about 15%. The current market potential of herbal medicine is estimated about $ 80–250 billion in Europe and the United States.

Medicinal plants throughout have played a pivotal role in the design and development of potent therapeutic agents. During 1950–1970, approximately 100 new plant-based drugs were introduced in the US drug market including vinblastine, vincristine, reserpine, deserpidine, and reseinnamine, which are derived from higher plants. From 1971 to 1990, new drugs such as Egug-gulsterone, nabilone, plaunotol, ectoposide, lectinan, teniposide, Zgug-gulsterone, ginkgolides, and artemisinin appeared all over the world. Merely 2% of plant-derived drugs were introduced from 1991 to 1995 including gomishin, irinotecan paciltaxel, toptecan, etc. Plant-derived drugs have provided exceptional contribution toward modern therapeutics, for example, in the year 1953, serpentine isolated from the root of Indian plant *Rauwolfia serpentina* was a revolutionary finding in the treatment of hypertension and lowering of blood pressure. Vinblastine, a major constituent of *C. rosesus*, is used for the treatment of leukemia in children, Hodgkins, testicular and neck cancer, choriocarcinoma, and non-Hodgkins lymphomas. Vincristine, on the other hand, is recommended for acute lymphocytic leukemia in childhood advanced stages of Hodgkins, cervical cancer, lymophosarcoma, and breast cancer. Phophyllotoxin derived from *Phodophyllum emodi* is currently used against testicular, small cell lung cancer, and lymphomas. Plant-derived drugs have also been reported to cure mental illness, jaundice, hypertension, skin diseases, tuberculosis, diabetes, and cancer. Thus, all the above facts do confer that medicinal plants indeed have played an important role in the development of potent therapeutic agents.

Plants shall continue to be used worldwide for the treatment of disease and as a source of novel drug entities through research at both the academic and industrial levels. Despite the massive arsenal of clinical candidates developed by the pharmaceutical industries, an aversion has been seen by many members of the modern society and thus making herbal remedies a popular alternative for the treatments and eradication of diseases worldwide.

FURTHER READING

Anonymous, 1948–1976. Wealth of India, vol. 1–11(CSR), New Delhi.

Cragg, G.M., Newman, D.J., Snader, K.M., 1997. Natural products in drug discovery and development. J. Nat. Prod. 60, 52–60.

Cragg, G.M., Newman, D.J., 2012. Natural products as sources of new drugs over the 30 years from 1981 to 2010. J. Nat. Prod. 75, 311–335.

Iwu, M.M., 1993. Handbook of African medicinal plants. CRC Press, Florida. p, 257.

Tan, P.V., Penlap, V.B., Nyasse, B., Nguemo, J.D.B., 2000. Anti-ulcer actions of the bark methanol extract of Voacanga africana in different experimental ulcer models in rats. J. Ethnopharmacol. 73, 423–428.

Chapter 4

What All Should Know about Plant Drugs

Chapter Outline

Essentials of Botanical Extraction. http://dx.doi.org/10.1016/B978-0-12-802325-9.00004-5

4.1 ROLE OF PLANTS IN DRUG DEVELOPMENT

4.1.1 Plant Secondary Metabolites as Potent Leads for Pharmacologically Active Compounds

Unfortunately, there have been very few industrial endeavors when it comes to drug discovery from plants. Perhaps the most prohibitive factor might be the enormous cost involved while isolating a compound and conducting further research on it. Indian has been lagging behind the Western countries when drug discovery from new chemical entities from plants is concerned. Following are the few examples whereby plant secondary metabolites have contributed in the development of life-saving therapeutic compounds.

4.1.1.1 Anti-Inflammatory and Analgesic Leads from Plants

The term inflammation is derived from the Latin word—*Inflammar.* Any form of injury to the human body can bring out a series of chemical changes in and around the injured area. Inflammation results in the liberation of endogenous mediators namely histamine, serotonin, bradykinin, prostaglandins etc. Most of the anti inflammatory drugs now available are potential inhibitors of cyclooxygenase (COX) pathway of arachidonic acid metabolism which produces prostaglandins. Today herbal drugs are routinely used for curing diseases rather than chemically derived drugs which are known to have profound side effects (indigestion, stomach ulcer, gastrointestinal bleeding, anemia, hypertension etc.). So it is very important that an exhaustive research with ethnobotanical plants possessing anti-inflammatory and analgesic properties be carried out which can definitely open up new dimensions in inflammatory disorders and related complications. Plant extracts or purified natural compounds from plants can serve as template for the synthesis of new generation anti-inflammatory drugs with low toxicity and higher therapeutic value. Following is the list (non exhaustive) of plant-derived phytocompounds with significant anti-inflammatory potential:

- Withanolides from *Withania somnifera* (Ashwagandha or Indian ginseng)— Withania is regarded as an anti-inflammatory herb in Ayurveda and has traditionally been used for treating arthritis and asthma. Withania exerts its positive effects on arthritis through a direct anti-inflammatory, antioxidant and chondroprotective effect. It is postulated that anti-inflammatory and antioxidant actions of Withania also combine to contribute to its antitumor properties as well. Withanolides, which are constituents of Withania, acts by directly suppressing the activation of nuclear factor kappa B (NF-κB) which is induced by a number of inflammatory and carcinogenic mediators including tumor necrosing factor (TNF) and interleukin 1β.

- Boswellic acid from gum resin of *Boswellia serrata*—Boswellic acids inhibits the leukotriene synthesis via 5-lipoxygenase, but did not affect the 12-lipoxygenase and the COX activities. Additionally, boswellic acids did not impair the peroxidation of arachidonic acid by iron and ascorbate. Thus boswellic acids are specific, nonredox inhibitors of leukotriene synthesis either interacting directly with 5-lipoxygenase or blocking its translocation.

- Berberine from *Berberis aristata*—Berberine suppresses proinflammatory responses through adenosine mono phosphate-activated protein kinase activation in macrophages. Berberine significantly down regulated the expression of proinflammatory genes such as TNF-α, interleukin-1beta (IL-1β), IL-6, monocyte chemoattractant protein-1, inducible nitric oxide synthase (iNOS), and COX-2.

- Curcumin from *Curcuma longa*—Curcumin down regulates various proinflammatory cytokine expressions such as TNF-α, interleukins (IL-1, IL-2, IL-6, IL-8, IL-12) and chemokines, most likely through inactivation of the nuclear transcription factor, NF-κB. Likewise, curcumin is also known to decrease the inflammation associated with experimental colitis, including a substantial reduction of the rise in myleoperoxidase activity, an established marker for inflammatory cells (mainly polymorphonuclear leukocytes) and TNF-α. In addition, curcumin is able to downregulate COX-2, iNOS expression and p38 mitogen activated protein kinase activation

- Gugglesterone—*Commiphora mukul*—Guggulsterone has been found to potently inhibit the activation of NF-κB, a critical regulator of inflammatory responses. Such repression of NF-κB activation by guggulsterone has been proposed as a mechanism for the anti-inflammatory effect of guggulsterone.

- Nimbidin from *Azadirachta indica*—Nimbidin inhibits nitric oxide (NO) and prostaglandin E_2 (PGE_2) production in lipopolysaccharide (LPS) stimulated macrophages following *in vitro* exposure, whereas interleukin 1 (IL-1) was only weakly inhibited. Probing the mechanism of NO inhibition revealed that nimbidin ameliorated the induction of inducible NO synthase (iNOS) without any inhibition in its catalytic activity.
- Bromelain from *Ananas comosus* (L.) Merril (Pineapple)—Bromelain selectively decrease thromboxane generation and change the ratio of thromboxane/prostacyclin in favor of prostacyclin rather than blocking the arachidonic acid cascade at the enzyme COX. Bromelain has been shown to inhibit prostaglandin even though its action is significantly weaker. The anti edema, anti-inflammatory and coagulation inhibiting effects are due to an enhancement of the serum fibrinolytic and fibrinogenolytic activity. It blocks synthesis and lowers serum and tissue levels of kinin compounds.
- Cannabinol, from *Cannabis sativa*—Traditional use of Cannabis as an analgesic, anti-asthmatic and antirheumatic drug is well established. Cannabinoids obtained from cannabis has been shown to inhibit PG mobilization and synthesis. Cannabinoids also stimulate and inhibit phospholipase A_2 activity as well as inducing an inhibition of COX and lipoxygenase.
- Betulinic acid from *Bacopa monnieri*—In Ayurvedic medicine, it is extensively used to treat various inflammatory conditions such as bronchitis, asthma and rheumatism. Its anti-inflammatory action is due to its ability to block the calcium channels which results in the suppression of prostaglandin E_1, bradykinin and serotonin secretion.
- Salicin from *Salix alba*—Salicin which is a phenolic glycoside of the Willow tree (*S. alba*) is known to inhibit the enzyme COX thus blocking the production of PG_2 and exerts a mild analgesic action.

Figure 4.1 depicts the mechanism of action of some anti-inflammatory leads from plants.

4.1.1.2 Cardiotonic Leads from Plants

Cardiotonic are drugs used to increase the efficiency and improve the contraction of the heart muscle, which leads to improved blood flow to all tissues of the body. Cardiotonic drugs increase the force of the contraction of the muscle (myocardium) of the heart. This is called a positive inotropic action. When the force of contraction of the myocardium is increased, the amount of blood leaving the left ventricle at the time of each contraction is increased. When the amount of blood leaving the left ventricle increases, cardiac output (the amount of blood leaving the left ventricle with each contraction) is also increased.

- Digoxins, digitoxin from *Digitalis purpurea*—The therapeutic uses of cardiac glycosides are primarily for two purposes. During heart failure digoxin increases the rate of contraction of heart muscle and the lowers edema. Whereas

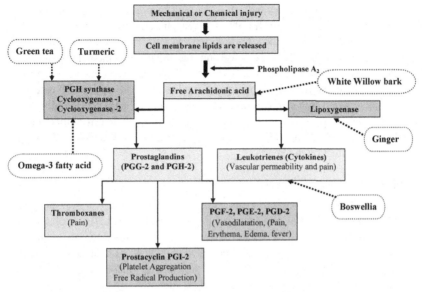

FIGURE 4.1   Anti-inflammatory and analgesic leads from plants.

during arrhythmias it decreases atrioventricular (AV) nodal conduction (i.e. it shows *para*-sympathomimetic effect) and decreases ventricular rate.

- Ouabain from *Strophanthus gratus* and *Acokanthera ouabaio*—It is used primarily as a research tool. It is a cardiac glycoside occurring in ripe seeds of *S. gratus* and the bark of *A. ouabaio*. Ouabain binds with Na^+—K^+ATPase pump and inhibits its action thereby causing a rise in intracellular sodium ions. The whole process actually enhances cardiac ionotropy and thus become useful as cardio tonics.
- Fruits of Yellow oleander plant (*Thevetia neritfolia*)—It is reported to contain thevitin A, B and peruvoside which are potent cardiac glycosides.
- Bark of *Terminalia arjuna*—It has been used for the treatment of symptoms similar to angina pictoris in the traditional Indian system of medicine. Arjunolic acid isolated from this plant has been found to possess significant cardiac protection.

4.1.1.3 Antihypertensive Leads

- *Reserpine*—It is an indole alkaloid obtained from *Rauwolfia serpentina*. It is used as antipsychotic and antihypertensive agent for the control of high blood pressure and for the relief of psychotic symptoms respectively.

  The antihypertensive action of reserpine is due to its ability to deplete the concentration of catecholamines from the peripheral sympathetic nerve endings. It irreversibly blocks the vesicular monoamine transporter which actually transports released and free nor-adrenaline or nor-epinephrine, 5-hydroxy tryptamine and dopamine from the cytoplasm of the presynaptic nerve terminal into storage vesicles for further release into the synaptic cleft as shown in Figure 4.2. Due to uptake blockade these unprotected and vulnerable neurotransmitters are metabolized by monoamino oxidase and catecholamine-*o*-methyl transferase in the cytoplasm and as a result never reach the synapse to produce their actions. Thus nor-adrenaline or nor-epinephrine mediated actions on α and β receptors (adrenoceptors) are terminated and a hypotensive state starts prevailing.

- The *Coleus* spp.—Coleonol (synonym: Forskolin) isolated from this plant at CDRI, Lucknow, has been shown to possess hypotensive action and positive inotropic effect on the heart.

4.1.1.4 Antidiabetic Leads

Diabetes mellitus is a group of metabolic diseases characterized by hyperglycemia resulting from defects of insulin secretion, insulin action, or both. The mechanism of action of insulin release from β-cells of pancreas in response of glucose in blood is a very complex process. Initially glucose enters the β-cells with the help of glucose transporter 2, which is phosphorylated by the enzyme glucokinase enzyme. The modified glucose is then metabolized to produce adenosine-triphosphate (ATP). The rise in ATP: ADP (adenosine-diphosphate) ratio causes the closure of ATP-gated potassium channels in the cell membrane thus preventing the flow of potassium ions. Due to this modification there is an increase in the internal positive charge of the cell causing it to depolarize. The net effect thus is the activation of voltage-gated calcium channels, which carries calcium ions into the cell and as a result triggers the export of the insulin stored granules from β-cells into the surrounding blood vessels by a process called exocytosis which enhances glucose intake, utilization and

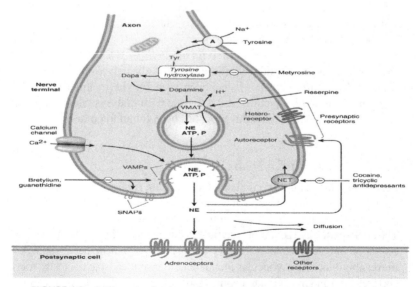

FIGURE 4.2 Antihypertensive drugs of natural origin acting on the presynaptic cleft.

storage in various tissues. In patients with diabetes, the body loses insulin producing capacity as a result of apoptosis of pancreatic β-cell or insulin insensitivity.

It's a well known fact that most of the glucose-lowering synthetic drugs (sulfonylureas, meglitinides, biguanides, metformin, thiazolidinediones, miglitol etc.) have side effects, including severe hypoglycemia, lactic acidosis, idiosyncratic liver cell injury, permanent neurological deficit, digestive discomfort, headache, dizziness and even death.

Medicinal plants have beneficial multiple activities including manipulating the carbohydrate metabolism by various mechanisms, preventing and restoring integrity and functioning of β-cells, insulin releasing activity, improving glucose uptake and utilization, and antioxidant properties. Several plants have been tested for their antidiabetic potential and in most of the cases the findings have been based on the ethno-botanical claims. The present nonexhaustive list gives an overview of some plants with well-known profiles of antidiabetic claims.

- Charantin and Vicine from *Momordica charantia*—It has been unanimously agreed that *M. charantia* fruit juice reduces blood glucose level. This is accomplished due to its effect on gluconeogenic enzyme and secondly by increasing intestinal Na^+/glucose co-transporters. The exact mechanism of its hypoglycemic effect is not yet established. However, it has been suggested that the depression of key gluconeogenic enzyme such as glucose-6-phosphatase and fructose—biphosphatase may be partly involved in the hypoglycemic effects of the fruit juice. Another opinion is that it may cause hypoglycemia effect via an increase in glucose oxidation through the activation of glucose metabolism and inhibition of glucose absorption from the gut. The probable mechanism of action of *M. charantia* as antidiabetic lead is shown in Figure 4.3.

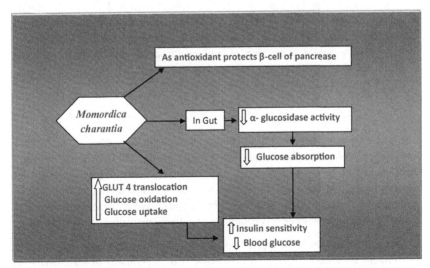

FIGURE 4.3 Mechanism of action of *Momordica charantia* as antidiabetic lead.

- Gymnemic acid from *Gymnema sylvestra*—It is a woody climbing plant indigenous to India. The leaves of this plant have been used in India over 2000 years to treat "*Madhumeha*" or "Honey urine". Its common name is "gurmar" or "sugar destroyer". Its antidiabetic activity appears to be by the combination of two mechanisms. Gymnema increases the activity of enzymes responsible for glucose uptake and utilization and on the other hand also inhibits peripheral utilization of glucose by somatotrophin and corticotrophin.

- Dietary fiber from *Trigonella foenugraecum*—Consuming fenugreek mixed with food (meal) seems to reduce postprandial blood glucose levels in patients with type I or type II diabetes. The applicable part of fenugreek is the seed. The active chemical components include trigonelline, 4-hydroxyisoleucine, and sotolon. Fenugreek seeds have a characteristic bitter taste and odor. Fenugreek seeds contain dietary fiber and pectin and may affect gastrointestinal transit of glucose and thus slowing glucose absorption. 4-Hydroxyisoleucine which constitutes about 80% of the total content of the free amino acids of the seed appears to be directly involved in stimulating insulin secretion. The mechanism of action is totally glucose dependent and only occurs in presence of moderate to high glucose concentration. In humans fenugreek seeds exert hypoglycemic effect by stimulating glucose dependent insulin secretion from pancreatic beta cells as well as by inhibiting the activities of α-amylase and sucrose-the two intestinal enzymes involved in carbohydrate metabolism.

- Allicin from *Allium sativum*—Allicin has been found to activate the enzymes hexokinase, glucose-6-phosphatase, 3-hydroxy-3-methyl-glutaryl (HMG) Co-A reductase and lecithin cholesterol acyl transferase. Moreover, inhibition of dipeptidyl peptidase-4 results in an increase insulin secretion and reduces glucagon secretion by preventing the inactivation of glucagon-like peptide-1, thereby lowering glucose levels.

- Ginsenosides from Ginseng species—Ginseng is able to decrease blood glucose by affecting various pathways possibly by blocking the intestinal glucose absorption and inhibiting the hepatic glucose-6-phosphatase action. In the gut it is found to inhibit the action of α-glucosidase activity and in the process decreases glucose absorption into the blood. Overall Ginsenosides from ginseng improves glucose metabolism.

4.1.1.5 Antimalarial Leads

A number of medicinal plants have been used traditionally for the treatment and management of malaria. Numerous life-saving Antimalarial drugs have been isolated from plants. Antimalarial drugs from plants have shown promising results however limitations such as toxicity, low bioavailability and poor solubility have restricted the scope of their use. Medicinal plants have provided valuable and clinically used antimalarials and notable among them are quinine and artemisinin.

- Quinine from *Cinchona* spp.—It is an early levo-rotatory alkaloid obtained from cinchona bark. Its D-isomer quinidine is used primarily as antiarrhythmic drug and sometimes as an antimalarial drug as well. Quinine mainly acts on the trophozoite stage of parasite development. It kills the sexual stages of *Plasmodium vivax, Plasmodium malariae, Plasmodium ovate* but not mature gametocytes of *Plasmodium falciparum*. Quinine is thought to be involved in the inhibition of parasitic "heme detoxification" in the food vacuole and does thought to be preventing hemozoin formation.

- Artemisinin from *Artemisia annua*—It is a sesquiterpene lactone extracted from the leaves of *A. annua*. It is active against all *Plasmodium* species. It acts at the early trophozoite stage and ring stages. It is active only on "blood-stage" of the parasite and have no effect on the "liver stage" of the parasite. Heme (with Fe^{2+}) produced during hemoglobin digestion of parasite, catalyses the opening of the peroxide bridge of Artemisinin leading to the formation of free oxygen radicals. Malarial parasites are prone towards these free radicals. It is generally termed as blood schizontocidal.
- Many biflavonoids like bilobetin and heveaflavone have been reported from *Selaginella bryopteris* which have been investigated for their antiprotozoal activity *in vitro* against the strain of *P. falciparum* and have been found to be potent.
- Neem is widely used in the traditional Indian system of medicine for a variety of complications. Nimbolide has been identified as the active antimalarial principle of this plant.

4.1.1.6 Antiobesity Leads

There are numerous remedies which are actually mentioned in Ayurveda and other traditional system of medicine practices for the treatment of obesity.

- Tea/*Thea sinenis*—Tea polyphenolics are the potent pancreatic lipase inhibitors. It contains (−) epi-gallocatechin and theaflavin. Pancreatic lipase actually assists in the formation of triglycerides whose excess deposition in the adipose tissues causes obesity.
- Guggulipid from *C. mukul*—Guggulipid which has been developed at CDRI, Lucknow, India as a hyperlipidemic agent actually acts by inhibiting the action of HMG-CoA reductase which catalyses the conversion of HMG-CoA into cholesterol and subsequently to bile acids. So if no cholesterol and bile acid gets synthesized then no triglyceride formation would take place and thus no fat deposition in the adipose tissues would occur.
- Other important antiobesity drugs from medicinal plants are carnosic acid from *Salvia officinalis,* crocin and crocetin from *Gardenia jasminoids.*

4.1.1.7 Antileishmanial Leads

Usefulness of natural products in drug discovery and its subsequent development into a potent therapeutic lead is not surprising at all as many of the medicinal plants like cinchona bark, strychnos bark, willow bark etc. were historically used against different diseases caused by protozoans and other parasites.

Botanicals have been used in traditional medicine for treating various types of cutaneous leishmaniasis. In leishmanial parasites, topoisomerase I is the enzyme that induces DNA strand break, manipulation and rejoining activities to directly modulate DNA arrangement necessary for parasite growth and multiplication.

- Flavonoids like luteolin and quercetin was found to inhibit the growth of *Leishmania donovani* promastigotes and amstigotes and induce cell cycle arrest leading to apoptosis.
- Diospyrin isolated from *Diospyros* spp. has been a very potent antileishmanial natural product against *L. donovani*. It is found to inhibit the type I DNA topoisomerase of *L. donovani* parasite.
- Plumbagin from *Plumbago* spp. is perhaps the most potent antileishmanial natural product with against *L. donovani*.
- Piperine has been shown to be active against promastigotes of *L. donovani* with activity comparable to pentamidine.
- Amarogentin, isolated from *Swertia chirata* has been found to inhibit *L. donovani* topoisomerase also.
- A standardized mixture of iridoid glycosides prepared from the root and rhizome extract of *Picororrhiza kurroa* popularly known as Picroliv shows a noteworthy antileishmanial activity and is used in therapy of *kala azar* in combination with sodium stibogluconate where it has been reported to enhance the efficacy of the antileishmanial drug and also to reduce its side effects.

4.1.1.8 Anticancer Leads

Plants have proved to be an important source of anticancer leads for several years. Almost about 30 plant derived compounds have been isolated so far and are currently under clinical trials. These anticancer leads have been found to be clinically active against various types of cancer cells. Different anticancer compounds that have been identified and reported by scientists have been reviewed under.

- Vinblastine and vincristine from *Catharanthus roseus*—These agents do not work to alter DNA structure or function. These drugs impede with the mechanics of cell division. During mitosis, the DNA of a cell is replicated and then gets divided into two new cells. This process involves spindle fibers, which are constructed with microtubules (formed by long chains of smaller subunits of proteins called tubulins). Vinca alkaloids are known to bind especially with the β-tubulins thus inhibiting the polymerization of the spindle fibers as shown in Figure 4.4. Without functional spindle fibers or spindles, the cell cannot divide and eventually dies out. So, the spindle inhibitor drugs function in a cell-cycle dependent manner.
- Podophyllotoxin from *Podophyllum peltatum*—The mode of action of Podophyllotoxin is that it binds to tubulin and is a member of spindle poison group of agents and functions by preventing microtubule formation. This natural compound has been used to generate two semisynthetic derivatives

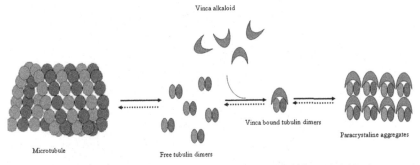

Vinca alkaloid

Microtubule

Free tubulin dimers

Vinca bound tubulin dimers

Paracrystaline aggregates

FIGURE 4.4 Mechanism of action of anticancer leads vincristine and vinblastine.

namely—etoposide and teniposide both of which work via different mechanisms by inhibiting the enzyme topoisomerase II thus preventing DNA synthesis and hence its replication as shown in Figure 4.5.

- Paclitaxel (Taxol®) from the bark of *Taxus brevifolia*—Paclitaxel-treated cells have defects in mitotic spindle assembly, cell division and chromosome segregation during mitosis. Unlike other tubulin-targeting drugs such as colchicines and vinca alkaloids that inhibit microtubule assembly, paclitaxel stabilizes the microtubules and protects it from depolymerization as shown in Figure 4.6. Chromosomes are thus unable to achieve a metaphase spindle configuration. This impede progression of mitosis, and a prolonged activation of the mitotic checkpoint results in apoptosis or degeneration to the G-phase of the cell cycle without cell division and the cell dies out eventually.

- Camptothecin from *Camptotheca acuminate*—A remarkable aspect of camptothecin is their specificity for binding to and inhibiting a precise step of the Topisomerase1 reaction. Camptothecin binds neither to Topisomerase1 alone nor to DNA but only to the complex formed by Topisomerase1 when it cleaves DNA. This denotes that Camptothecin selectively binds to the cleavage DNA complex. This prevents DNA re-ligation and therefore causes DNA damage which results in apoptosis. Mechanism of action of anticancer lead Camptothecin and its analogs is shown in Figure 4.7.

4.1.2 Plant Derived Extracts Currently under Clinical Trials

Many registered plant derived medicines which are not single chemical entities have often been subjected to quality control via extract standardization procedure involving biomarker compounds with known or unknown biological activity. Following are a few examples of some standardized plant extracts which have gone under clinical trials for the treatment of several diseases.

- Devil's claw or *Harpagophylum procumbens*—It is a South African plant that has been used for years in Africa for the treatment of arthritis and fever. The active ingredients of Devil's claw are harpagoside, harpagide and procumbine.

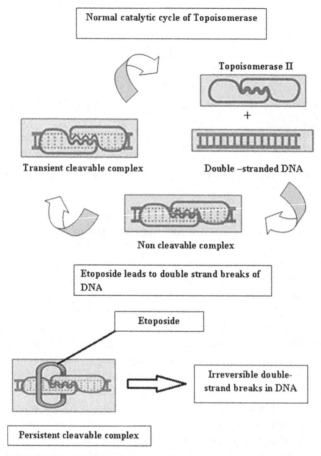

FIGURE 4.5 Mechanism of action of anticancer lead podophyllotoxin.

Harpagoside has been shown to suppress LPS induced inducible NOS and COX-2 expression through inhibition of NF-κβ activation. This Devil's claw extract is currently undergoing Phase 2 clinical trial for the treatment of hip and knee osteoarthritis.

- Flavocoxid—It is a blend of natural flavonoids from *Scutellaria baicalensis* and *Acacia catechu*. It is generally used as medicinal food for the management of osteoarthritis. It works by inhibiting COX-1 and COX-2 and lipo-oxygenase enzyme system. It is currently undergoing Phase 1 clinical trial for the treatment of knee osteoarthritis.
- Ginkgo extract—This is obtained from the dried leaves of *Ginkgo biloba* tree often described as a living fossil with no close living relatives. Standardized ginkgo extract contains flavone glycosides (quercetin, kaempferol and iso-hamnetin) and terpenoid lactones. It is used in the treatment of early stages

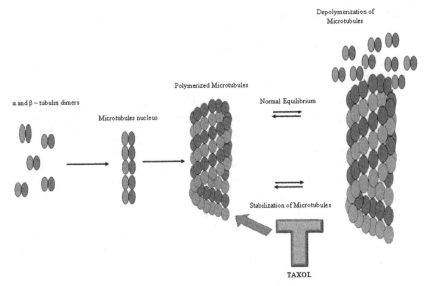

FIGURE 4.6 Mechanism of action of anticancer lead paclitaxel.

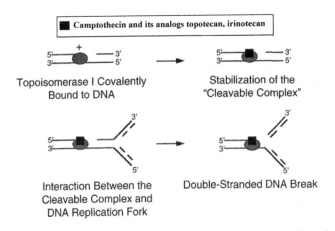

FIGURE 4.7 Mechanism of action of anticancer lead camptothecin and its analogs.

Alzheimer's disease and in vascular dementia. It is currently undergoing Phase 3 clinical trial to prevent dementia and the onset of Alzheimer's disease.

- Mistletoe (*Viscum album* L.)—Mistletoe is a semiparasitic plant that grows on several types of common trees such as Apple, Oak and Pine. It is one of the most widely studied complementary and alternative medicine therapies for the treatment and management of cancer. In USA, mistletoe extracts is currently undergoing Phase 2 clinical trial for the treatment of lung cancer in patients receiving conventional chemotherapy and in combination with gemcitabine (a synthetic antitumor drug) for the patients with advanced solid tumor. The extract contains a cytotoxic lectin and various cytotoxic proteins that have been shown to induce tumor necrosis, macrophages activation, apoptosis induction apart from protecting DNA in normal cells during conventional chemotherapy.

All the above cited examples of medicinal plants do possess many novel prototypes of active molecules, some of which have led to life-saving valuable drugs which are available on the market today and also as potent candidates for clinical trials. Moreover, modification of extracts obtained from medicinal plants into a chemically engineered extract can be an alternative source of bioactive compounds and has become a common technique which is being utilized by various synthetic and natural product chemists.

During the last three decades there has been a remarkable achievement in the area of natural product-based drug discovery where several bioactive compounds with profound bioactivity have been discovered with the aid of modern and state of art sophisticated techniques.

4.2 FACTORS THOUGHT TO BE THE REASON FOR DECLINING INTEREST IN BOTANICALS

Despite these past successes and advantages many large pharmaceutical firms have decreased the use of natural products in new drug screening and discovery during the early 90s.

Reasons thought to be responsible for declining natural product driven drug discovery are as follows:

4.2.1 Ethical Issues and Related Complications

Concerns regarding the loss of biodiversity and legal formalities with relation to medicinal plant procurement and use can be time-consuming. Stringent laws to prevent biopiracy were undertaken in the Earth Summit, Rio de Janeiro, Brazil (1992) where it has been asserted that countries have sovereign rights to protect and preserve its own bio diversity. Under such circumstances research involving medicinal plants need to be carried out within an agreed legal framework. There is a fear of widespread destruction of the ecosystems if bioprospecting is allowed to flourish as it will threaten to eliminate many known and unknown plant species. Overall such mindset would actually prevent the scientific exploration and exploitation of thousands of species whose medicinal potential will never be known to the world.

4.2.2 Nonavailability of Materials According to Market Demands

Concern over the availability of required quantity can be regarded as one of the main causes of declining interests in botanical-based drug discovery among the pharmaceutical industries. A large quantity is required initially for the generation of information to understand and assess the actual potential of the substance understudy especially with respect to preclinical studies.

For plant derived bioactives physical access for a recollection may be difficult, or permission to collect and shipment of the plant material may be hard to obtain. In addition to these issues plants are also known to produce quantities of desired bioactive compounds under certain environmental or ecological conditions. Moreover, the market need or demand can reach a scale of hundreds to thousands of kilograms per year if the bioactive(s) under investigation turns out to be a potent one.

4.2.3 Difficulties in Obtaining Authenticated Plant Materials

It is often quite easy to collect plants with notable medicinal values and to demonstrate their biological potentials in labs. It has been a mindboggling issue with researchers to find the same plant every time they visit the collection site due to improper documentation and unscientific plant restoration programs. The use of a wrongly identified plant is quite common and accidents due to botanical misidentification seldom occur. For example, Brazilian folk medicine uses a plant called *Quebra-pedra* as diuretic and also for the treatment of gallstone problems. But the correct plant is *Phyllanthus nirurii* which is commonly confused with species from *Euphorbia* genus, which are known to be very toxic in nature.

4.2.4 Problems Associated with the Measurement of Biological Activities of Extracts

Botanicals and its preparation are complex mixtures of several materials. Several types of interactions among the components of the mixtures, either antagonism of one component's bioactivity by another component or emergence of synergy between them often give very misleading results. Moreover, purification and identification of active constituents from these complex natural product mixtures having dozens to hundreds of different chemical substances of similar chemical and physical properties are often quite slow and are not cost-effective.

4.2.5 Time Constraints

Most of the pharmaceutical firms during the early 1990s were not interested in the evaluation of extracts obtained from plants. One of the reasons behind this is that when an extract shows activity in a particular type of bioassay understudy, under such circumstances the main aim then is to isolate and characterize the active principle responsible for the bioactivity. This particular last step is pretty expensive and may take a long time depending upon the availability of an appropriate amount of plant material, time required to perform a bioactivity and last of all is the structure elucidation of the bioactive(s) found in the extract.

4.2.6 Lack of Reproducibility of Bioactivity

It has been observed that the bioactivities which are detected during preliminary screening programs often do not repeat when plants are resampled and extracted. This ambiguity may be due to the fact that biochemical profiles of plants harvested and collected at different time and location varies greatly. Moreover, complex nature of plant extracts complicates the task of potency evaluation and novelty of the bioactive ingredient which is present in trace quantity.

4.2.7 Natural Products Obtained from Plants Lack Desired Biophysical Properties

In order to produce a better synthetic compound the synthetic chemists favours compounds which have better biophysical properties so that better orally active drug candidates can emerge. Thus, compounds should be under molecular weight of 500 Da, posses less than 5 hydrogen bond donors, less than 10 hydrogen bond acceptors, and have log P < 5. This is known as Lipinski rules of five. Lipinski's rule of five describes the properties and structural features that make molecules more or less "drug-like". This rule initially was thought to be not applicable to natural products obtained from botanicals as the size of most of the bioactives are more than 500 Da (bioactive compounds with molecular weights much beyond 500 Da cannot cross cell membranes and therefore cannot be made orally bioavailable) with more numbers of hydrogen bond donors and acceptors.

4.2.8 Problem of Rediscovering of an Already Known Compound

Rediscovering of known compounds is a main obstacle when screening of botanical extracts is performed. This occurs due to an inapt de-replication methodology which is often practiced. On the other hand, the dereplication process is quite a time-consuming process and the methodologies thus adopted are not at par with the modern regime of "Blitzkrieg" screening methods.

4.2.9 Discovery and Development of Natural Products from Botanicals is Considered to be a Slow Process and Does Not Match the Pace of High throughput screening

During the early 1990s, major Pharma companies worldwide tilted towards a new lead finding strategy based on high throughput screening (HTS) of a very large collections or libraries of synthetic compounds. Prompt interest in HTS arose due to believe that techniques such as combinatorial chemistry and computational drug design could produce larger and more effective libraries with improved lead hit rates by maintaining quality at the same time. Moreover, advancements in molecular biology, cellular biology and genomics have dramatically increased the number of molecular targets thus expediting the need for shorter timelines in drug discovery. In other way it can be said that in today's drug discovery scenario rapid screening and quick identification of potential drug molecules is absolutely essential for success. Once a hit has been confirmed in biological screening program, the extract must be fractionated to isolate the active lead, and this process normally requires that bioassays be conducted at each level of purification. This put traditional medicinal driven drug discovery process to a standstill as it is based on a long and strenuous process of crude extract screening, bioassay guided isolation strategy followed by structure elucidation.

4.3 APPROACHES AND STRATEGIES TO IMPROVE THE STATUS OF DRUG DISCOVERY FROM BOTANICALS

While natural product drug discovery from botanicals suffers from lack of pharmaceutical industry support, several technological developments and advancements over the past several years are reducing the complexity of working and building these extracts thus making the field of natural product drug discovery more attractive and promising.

Several other alternative approaches are also being investigated in efforts to increase the speed and efficiency with which natural products can be applied to drug discovery.

4.3.1 Adoption of the Prefractionation Technique

Screening of plant extracts are sometimes complicated due to some nonspecific interferences arising from the presence of fluorescent, insoluble compounds and unwanted components like lipids, plant pigments, vegetable tannins etc. In addition, extracts are typically a complex mixture and the concentrations of active constituents in crude extracts usually remains unknown. Thus the throughputs can be lower when screening of synthetic chemical libraries is compared. However, advances in detection technologies and new screening assay techniques have overcome many of these challenges. Now hits are identified by various bioactivity assays in vast majority of industrial natural product discovery programs. In such programs the incompatibility of crude extracts with high throughput assays can be addressed by some degree of prefractionation of extracts. A modified betulinic acid derivative having HIV maturation inhibitor action has been discovered by this approach and the compound is currently in Phase 2 clinical trials.

4.3.2 Screening Assay Techniques

4.3.2.1 Whole Cell-Based Assay Method

It helps in the assessment of specific molecular interactions between plant extract components and living cellular system. Moreover, information about drug penetration and binding with receptors are obtained at a very early stage from such studies. It encompasses various approaches ranging from simple growth inhibition assays whereby the spectrophotometric or turbidimetric method for detection of activity is used. Sometimes a red blood cell lysis assay or an antiyeast growth assay may be employed to assess various cytotoxicity or nonspecific responses of plant extracts.

4.3.2.2 Biochemical Assay

When acquiring of target specific information is of prime concern then biochemical assay is the technique of choice. Various newer biochemical assay techniques now have come to the forefront such as capillary electrophoresis technique, enzyme-linked immunosorbent assays, high-throughput pharmacological screening and high-throughput mass spectrometry based screening assay.

Thus bioactivity assays techniques can be employed to identify lead structures within a short span of time.

4.3.3 Differential Smart Screening Technique for Activity Profiling of Extracts

Technological advancements achieved in separation and purification science have made natural product drug discovery from medicinal plants more compatible with the expected timescale of HTS techniques. Several alternative approaches are also being explored in order to increase the speed and efficiency with which botanicals from natural products can be applied to drug discovery process. Natural product drug discovery from botanicals is seeing a renewed interest due to the application and incorporation of high throughput pharmacological screening techniques. One such technique is popularly known as differential smart screen whereby the various ranges of bioactivity that is anticipated to be present in a crude extract can be easily measured with respect to several closely related receptor subtypes and the bioactivities are then matched with that of known compounds. The matching patterns thus obtained will suggest that the extract do require further investigation. The potential of differential smart screening technique can be enhanced upon coupling with various genomic and genetic manipulation techniques applied in plant tissue culture.

4.3.4 Chemically Engineering an Extract

Chemically engineering an extract is a novel strategy to generate bioactive compounds through chemical modification of inactive components present in extracts. This process focuses on the transformation of chemical groups highly common in natural products into chemical groups that are rarely produced by the secondary metabolism. Chemically engineered extracts can become an alternative source of compounds to feed the drug discovery process for new molecules with interesting biomolecular activities.

4.3.5 Amalgamation of In silico Screening Approach with Natural Products

Secondary metabolites obtained from plants are usually associated with improved nutritive value and have beneficial effects on human beings and animals. However, mechanism of toxicity and health promoting effects of most of the plant secondary metabolites are not well established. In recent years interest in plant secondary metabolites has risen dramatically among the plant molecular biologists, medicinal chemists and phytochemists owing to many reasons:

- Advancements in phytochemical structural analysis.
- Ease to synthesize the phytoconstituents along with their modification in order to suppress or ameliorate certain characteristics namely solubility, efficiency, or stability for better bio-activity and reduced toxicity.
- Structural modifications using in silico approach are quite significant in terms of money and time

A schematic depiction of in silico screening approach in natural product drug discovery is shown in Figure 4.8.

In order for the botanical drug discovery to continue to be successful, new and innovative approaches are absolutely necessary. By applying the previously mentioned approaches in a systematic manner it may be possible to increase the current efficiency in identifying and developing new drug candidates from botanicals. It can be asserted that by coupling these approaches with advanced extraction technologies, the time line for screening and discovery of novel drug leads from mother nature can be shortened.

4.4 APPROACHES IN MEDICINAL PLANT SELECTION PRIOR TO EXTRACTION

A correct choice of the vegetable matter has to be made when the aim is to isolate and identify bioactive substances. Various approaches which were used and can be considered now are presented below:

4.4.1 Ethnobotany and Ethnopharmacology Approach

Present day drug screening from plants find its root in the medicinal folklore which has evolved over the years and has proved to be an invaluable guide for

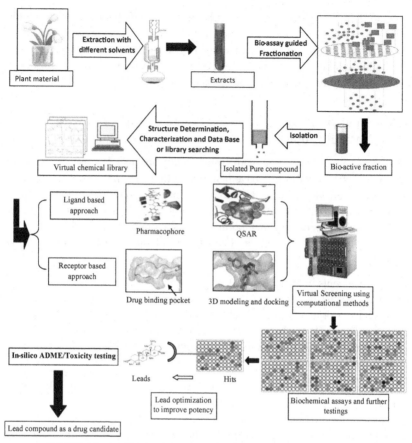

FIGURE 4.8   In silico screening approach in natural product drug discovery.

most of the life saving drug leads like digitoxin from *D. purpurea*, reserpine from *R. serpentine* tubocurarine from *Chondrodendron tomentosum*, ephedrine from various genus of *Ephedra*, ergonovine (synonymous: ergometrine) from *Ipomea violaceae*, atropine from *Atropa belladonna*, vincristine from *C. roseus* and aspirin from *Salix* species. The term "Ethnobotany" which is a subfield of ethnological as well as botanical sciences investigates the relationship between ethnic groups and their herbal environment including plants used as food, medicine, and raw materials. The term "Ethnomedicine" is a subfield of ethnobotany and studies the pharmacopoeia of an ethnic groups transmitted orally from generation to generation. "Ethnopharmacology" eventually investigates the pharmacological mechanism of plants used in traditional medicine by indigenous group. An Ethnopharmacology is one of the most useful approaches among the three as it facilitates a targeted search. The concept of Ethnopharmacology was initiated by the missionaries (like the Jesuits in sixteenth century, Latin America) who were interested in the use of pharmacologically active plants where ever they went. A molecular aspect to this concept was later established by various scientists like Luis Lewin [1850–1929], Carl Hartwich [1851–1917], Alexander Tschirch [1856–1939] and Richard Evans Schultes [1915–2001], who contributed in the study of the chemistry of pharmacologically active plants.

In fact there are many examples of pharmaceutically relevant substances, which were developed based on ethnopharmacological approach. Milestone discoveries are as follows:

- Elucidation of the pharmacological principles in foxglove (*Digitalis* spp.),
- Poppy from *Papaver somniferum*.
- Curare from *C. tomentosum* and *Strychnos* spp.
- Antimalarial quinine from various *Cinchona* spp.
- Tobacco as a stimulant was first observed in Latin America by the early colonialists and the seeds of *Nicotiana tabacum* L. where brought to Europe from Brazil in 1560 by Jean Nicot de Villemain. Nicotine was isolated and its structure determined in the nineteenth century.
- Aspirin was developed based on ethnopharmacological studies with the bark of the willow tree (*Salix* spp.), which has been used traditionally in Europe to treat fever and inflammation.
- Chinese antimalarial plant *Artemisia annua* L., which has resulted in the recent development of artemisinin, a sesquiterpene with a trioxane peroxide bond, into the new clinical semisynthetic antimalarial agent artemether.
- Reserpine from traditional medicinal plant *R. serpentine*. It is known as "Chota chand" in Hindi by the local people of Himalayan mountains for snake bite. Local people claim that in ancient times mongooses were observed to feed on the plant before engaging in a combat with cobra. Although many plants may not neutralize the venom itself, they may be used to treat snake bite because they alleviate some of the symptoms (fear and panic) by tranquilizing compounds, e.g., *R. serpentina*—"Sarpagandhi" (Apocynaceae) since this plant contains the tranquilizing alkaloid, Reserpine.

Ethno-directed approach to traditional knowledge has been extremely useful in screening and identification of plants with bioactive compounds with potential application in drug development. The ethno-directed approach has significantly increased the chances of discovery of new bio-molecules with potential therapeutic application while reducing the cost and time involved in this process. Most of these compounds are part of routinely used traditional medicines and hence their tolerance and safety are relatively better known than any other chemical entities that are new for human use. Thus medicinal plant search based on ethno-directed approach will offer an unmatched structural variety in delivering promising new leads.

4.4.2 Chemotaxonomic (Phylogenetic Approach) Approach

Chemotaxonomic approach can be employed to target a specific taxonomic group of medicinal plants containing specific compounds that are similar to those present in other species or genera which have previously exhibited a high "HIT" rates for a particular type of bioactivity. The occurrence of different compounds that can be used as biosynthetic markers is used by several botanists in taxonomical related studies and such approach can be used as a successful tool in the assortment of families, subfamilies and genera to be investigated in terms of produced metabolites. Thus in summary it can be asserted that through chemotaxonomy approach, one can select plants from known families and genera to produce certain classes of substances (e.g. glycosides, terpenoides, alkaloids, flavonoids, steroids, etc.), especially those recognized by their biological activities and therapeutic applications.

4.4.3 Random Approach

The randomized investigations consist in random selection and collection of plant species for study, according to the plant availability in a particular region. When carried out in regions with high diversity the likelihood of discovery new substances, bioactive or not, is certainly higher in this type of approach. It is an indispensable approach, once it can demonstrate the potential of different plant species that had never been investigated. This type of selection provides an endless source of new structures, since nature is an endless source of chemical library.

Random search is considered extremely laborious and yields success rates in the order of one new product per 10,000 screened plants. Between 1960 and 1981, the National Cancer Institute in collaboration with the United States Department of Agriculture collected and tested more than 114,000 extracts of some 35,000 plants against a range of animal tumor systems, essentially cell cultures. Nonetheless, important drugs have been discovered using this method in the likes of taxol, derivatives of camptothecin and homoharringtonine.

4.4.4 Ecological Approach or Field Observations

Ecological observation is another way to obtain bioactive leads from plant. It encompasses observation of interactions between organisms with their ecological environment. The absence of predation in an area infested with herbivores can indicate the presence of toxic compounds in the plants. Extraction of toxic compounds for medicinal use can be done from such areas. Metabolites involved in plant defense against microbial organisms may be useful as antimicrobial agents in humans, since they are not very toxic.

Likewise, secondary products with repellent action (e.g. unpleasant flavor or odor) against herbivores by neurotoxic activity could have beneficial effects in humans in the form of sedatives, antidepressants, anesthetics, muscle relaxants through their action on the CNS. Therefore, the astute observation of these eco-logical relationships is a useful tool in the selection of plant species.

4.4.5 Zoopharmacognosy Approach

A variation of the ecological approach is the Zoopharmacognosy. The notion that animals can self-medicate, when they have access to a variety of herbs, has fascinated mankind for years. The concept of self-medication in nonhuman ver-tebrates can be taken into consideration while selecting plants when the aim is to find a probable "HIT". Zoopharmacognosy describes a process by which wild animals select and use plants possessing medicinal attributes for their self medi-cation against diseases and parasites. Following examples of animal's self-medi-cation can be considered while selecting candidate plants for new drug discovery:

- Clues for plants containing stimulants—*Croton megalobotrys* (Euphorbiaceae), *Euphorbia avasmontana* (Euphorbiaceae), *Datura innoxia* and *Datura stra-monium* (Solanaceae) are among the few plants on which the Chacma baboons of South Africa are known to feed on an occasional basis. These plants are not under the regular diet of the baboons, instead are classified as "euphorics".
- Clues for plants with antischistosomal "HITS"—In Ethiopia, anubis baboons are found to feed themselves on the fruits and leaves of desert date (*Balanites aegyptica*) to control schistosomiasis—a parasitic disease caused by flat-worms of the genus schistosoma. Later scientific studies revealed the presence of diosgenin, a hormone precursor with an anti-schistosomal property.
  Chimpanzees of Tanzania were also found to chew on the leaves of *Vernonia amygdalina*. Phytochemical analysis revealed that the bitter pith of *V. amyg-dalina* contains seven steroid glucosides which are capable of killing parasites that cause schistosomiasis, malaria and leishmaniasis
- Clues for plants with antimicrobial "HITS"—Sick chimpanzees (*Pan troglo-dytes*) in the forests of Tanzania are known to fold and swallow whole leaves of *Aspilia africana*. The physical irritation produced by the bristly leaves on an empty stomach increases gut motility and secretion resulting in diarrhea. This sheds the body of parasitic worms, a major cause of illness in chimps. Dusky-footed wood rats of California engage themselves in nest fumigation behavior to control fleas, ticks and mites. Animals that rest in nests or burrows are particularly susceptible to nest-borne parasites that carry disease. In 2002, researchers from Vassar College in New York showed that dusky-footed wood rats place bay leaves in their sleeping nests and regularly tears them to release fumigating vapors, significantly reducing parasite survival.

4.5 AN OVERVIEW ON PRE-EXTRACTION TECHNIQUES

A flowchart representing the steps involved in pre-extraction techniques before performing an extraction is given below for better understanding (Figure 4.9).

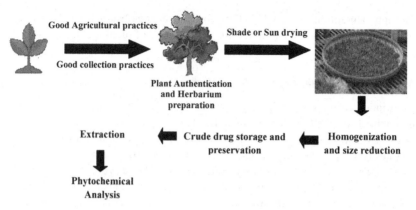

FIGURE 4.9 An overview on pre-extraction techniques.

4.5.1 Plant Sample Collection and Harvesting

The collection of wild plants is still very important task before initiating an extraction. Unfortunately, the intensive collection of wild plants often leads to excessive destruction of some species. In comparison with the harvesting of wild plants, cultivation has many advantages. It allows the collection of large quantities of raw materials from a relatively small area. Moreover, it means that the plant material has similar morphological characteristics and similar content of active components. Cultivation provides numerous benefits, providing, as a result of a single origin of the seed and the conditions of plantation, the raw materials are more balanced, both in terms of development and chemical composition.

The quantitative composition of a plant's active components depends on the time of harvest (the season and even the time of day), as the content of these substances is not constant throughout the year. Plant samples are collected from the aerial parts (e.g., herbs, leaves, fruits, seeds, stems, and stem bark), the trunk bark, and roots.

Usually, leaves are collected just as the flowers are beginning to open, flowers are collected just before they are fully expanded, and the underground elements are gathered as the aerial parts die down. Moreover, aerial parts of plants (leaves, flowers, fruits) should not be collected when they are covered with dew or rain. Plants that are discolored or have been attacked by insects or slugs also should not be collected.

When collecting plants in the field for natural product extractions, it is important to be properly prepared. Hereby we present some general guidelines.

- Wear field clothes and cover yourself head to toe if collecting is to be done in the cold winter or when mosquitoes or flies are in abundance
- Take along a note pad and pencil to record information about the collection site, soil conditions, ecological habitat, date of collection, plant identity, and who collected the plant(s)

- A pocket size field guide map and a hand lens can be an effective kit to help in proper plant identification
- If live plants are to be collected then Zip-lock plastic bags of various sizes in which the samples after collection can be kept should be used. Slips of adhesive papers are good to have for notes on plant identity regarding the collected specimen.
- Soil samples must be taken from the site of collection to get information on soil nutrient, soil pH, and soil type where such plants grows
- When collecting plants for extracts, it is important to get representative samples of all parts available: roots, vegetative shoots, bark from stems (if woody plant), flowers, fruits, and seeds (if mature).
- When collecting plants in the field, every last plant in the population should not be collected. This is important especially if the plant is rare, threatened or endangered.
- In the process of collecting herbaceous perennial plants, some original plants must be left intact where it is growing so that it can reproduce during the current and following years. Many of these plants take years to produce even a small amount of new biomass.

4.5.2 Plant Authentication and Herbarium Preparation

In phytochemical research, establishing the taxonomical identity of the plants at the initial stage of the research is a critical factor which is often overseen by researchers. A huge number of errors over establishing plant identity have been committed in the past and hence it is essential to authenticate the plant material whenever reporting new substances from plants or even known substances from new unexplored plants.

Once collected plant parts should be prepared and each should be authenticated by a competent taxonomist. The samples should be deposited in a recognized herbarium, so that future reference can be made to the plant studied if necessary. Details of the plant's origin, altitude of occurrence, climatic and microenvironmental conditions and any other allied information such as its local uses, overall health of the plant specimen and any other facts that may be useful for future investigation must be note down on a note card affixed with the voucher specimen. Above information is of vital importance in those cases where a recollection of the plant material is necessary and is beneficial for researchers to reproduce their work in future.

4.5.3 Plant Sample Drying

Overall, plants may be dried before extraction. However it is very important to ensure that drying operation should be performed under controlled conditions to prevent any kind of chemical deteriorations. The term "drying" should not be misinterpreted in the sense that it reflects total removal of moisture content.

Plant tissues do always contain some trapped moisture within the cell wall and hence the term dry mass means that the material may still contain 2–3% moisture. It is always advisable to report the total moisture content in terms of % w/w, the same has also been recommended in the WHO monograph developed for standardization of medicinal plants. Drying should be performed immediately after plant collection at ambient temperature in ventilated ovens with or without the flow of warm air or, sometimes nitrogen as per the requirements. Natural plant materials that contain volatile compounds—for example, essential oils, esters, vitamins, enzymes, and hormones—are particularly sensitive to higher drying temperatures. In these cases a drying temperature of 30–35 °C is required. Sometimes gradual drying can be applied, for example, first at a temperature of 20–30 °C and then at 80–100 °C (for cardiac glycosides). Connective ovens or vacuum ovens with water absorption, adsorption systems are often used to meet lab scale requirements. Effective drying can be performed at 40–50 °C; such low temperature drying prevents loss of volatile oil or natural fragrance of the plant material during drying. It is however, a more general practice to leave the sample to dry on trays at ambient temperature and in a room with adequate ventilation. Dry conditions are essential to prevent microbial fermentation and subsequent degradation of metabolites. Plant materials should be sliced or grind into small pieces and distributed evenly to facilitate homogenous drying. This grinding process in fact help during the extraction process whereby it assists in proper penetration of the solvent into the cellular structures of the plant tissues thereby assisting in dissolving the secondary metabolites and increasing extraction yield.

4.5.4 Size Reduction and Homogenization

The next operation in preparing raw plant material for analysis is proper size reduction or comminution and homogenization. As extraction is governed by the golden rule "decreased particle size—better surface area—better solute solvent contact—improved extraction" so proper size reduction is a very critical pre-extraction step. The choice of comminution technique depends on the consistency of the material, hardness and nature of constituents present. As in the case of plant material with volatile oil, temperature during comminution is a vital parameter. In such cases it is advisable that comminution must be carried out in small batch size. Harder plant parts like seeds, fruits are first cut manually and then ground in a mechanical mill. Since during manual cutting fragments of different sizes are produced, hence it is advisable to pass the material through a proper sieve. However, generally particles passed through sieve number 8 or sieve number 10 (as per Indian Pharmacopoeia.) is used for extraction and are referred to as coarse particles.

After size reduction homogenization of the plant material is usually carried out with ceramic mortar and pestle sets. A variety of mechanical homogenizers are also employed, but their use can result in local overheating of the material.

High energy ultrasonic vibrations, freezing under conditions resulting in the rupture of cellular wall, enzymatic breakdown, and other nonmechanical physicochemical process can also be employed for sample homogenization. Such processes are aimed to prevent any thermal loss of phytoanalytes.

4.5.5 Storage

Storage after collection is also a factor worthy of study. Appropriate storage of plant material is essential to prevent loss of bioactivity of the desired active compounds. Storage can also influence the physical appearance and chemical quality of plant materials and hence it is necessary to maintain appropriate storage conditions so as to increase their shelf life.

A basic condition for proper storage is the placing of plant material in a dry room at a low and controlled temperature. Presence of air humidity has a significant influence on the storage of plant material. Plants containing cardiac glycosides lose more than 50% of their bioactivity within one week, when the air humidity is 80%. Alkaloid and glycoside plant materials usually lose a few percent of their active ingredients in 5 or 10 years. Alkaloid plant materials are fairly stable, except for tropane alkaloids, which change during the drying process. Tannin plant materials lose 20% of their biological activity in 2 years as they are susceptible to oxidation, so they must not be stored in the powdered form. Flavonoids in raw plant materials are sensitive to the light, oxidation, and hydrolysis reactions, and oils in plant materials decay under the influence of higher temperatures. The time and conditions of plant material storage and humidity limits for individual plant materials are usually specified in several pharmacopoeias.

Upon storage plant materials do often get infected with pests. Fumigation on a small scale is often done with chloroform vapor but on a large scale tetrachloride carbon, ethyl bromide, or dichloroethane are used. However, these compounds are toxic, and they must be removed from plant material by proper ventilation.

FURTHER READING

Bhutani, K.K., Gohil, V.M., March 2010. Natural products drug discovery research in India: status and appraisal. Ind. J. Exp. Biol. 48, 199–207.

Chapter 5

Extraction of Botanicals

Chapter Outline

5.1 INTRODUCTION

The first step in the isolation and analysis of secondary metabolites is extraction which separates out the compounds from the cellular matrix. A fundamental knowledge about their place of occurrence, nature and characteristics is very much essential in selecting an extraction method. When the chemical nature of the secondary metabolite of interest is known to us an extraction method should be conducted in highly coordinated manner so as to obtain a high yield and purity. When the chemical composition is

Essentials of Botanical Extraction. http://dx.doi.org/10.1016/B978-0-12-802325-9.00005-7
Copyright © 2015 Elsevier Inc. All rights reserved.

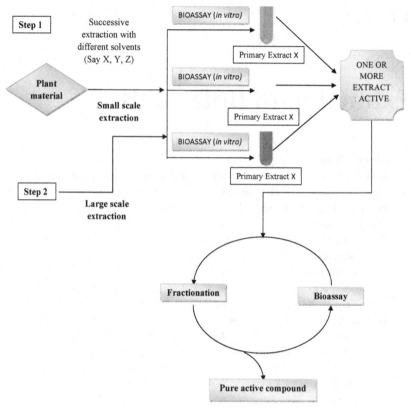

FIGURE 5.1 An example of natural botanical drug discovery process.

unknown it is worthwhile to reproduce the extraction methods employed traditionally (if reported) in order to enhance the chances of isolating potential bioactive metabolites.

An initial extraction is always performed typically on a small amount of plant material to obtain a *primary extract* as shown in Figure 5.1 below. This is done as part of a pharmacological study in order to gain preliminary knowledge on the nature of the extract and amount of secondary metabolites present in the plant material. Once specific metabolites have been identified in the initial extract, it may then become desirable to isolate them in larger quantities. This will involve recollecting a larger amount of plant material followed by a bulk or large-scale extraction.

Overall the aims of extraction process are as follows:

- To determine the composition of the natural product under investigation. This can be achieved either by physical assay method (thin layer chromatography, high performance thin layer chromatography, high performance liquid chromatography, liquid chromatography–mass spectrometry, liquid chroma-

tography–nuclear magnetic resonance) by comparing with some standard bio-
markers or by applying bioassay methods.
- To characterize the natural product by carrying out detailed experimental work
  by adopting various screening techniques.

Once a natural product source is selected or obtained, the first processing
step before further separation, identification, and characterization of bioactive
compounds is extract preparation. The three most fundamental questions that
should be asked at the outset of an extraction are:

1. What one is trying to isolate after an extraction?
   a. Is it an unknown compound(s) responsible for a particular biological
      activity?
   b. Is it a known compound(s) produced by a particular class of organism?
2. Why one is trying to isolate bioactive compound(s)?
   a. Is it to purify sufficient amount of unknown compound to characterize it
      partially or fully?
   b. Is it to fetch large supply of an already known compound so that exhaus-
      tive biological testing can be performed?
   c. Is it to determine what should be the required level of purity for biologi-
      cal testing, pharmacological testing and chemical profiling?
3. What is the purpose of performing an extraction process?
   a. Is it to scientifically validate the use of traditional medicinal plants which
      have been under use by our ancestors and traditional medicine practitio-
      ners for centuries?
   b. Is it to find an appropriate technique having resemblance with the tech-
      niques which were being used and practiced by the traditional medicine
      practitioner in order to increase the chances of finding new leads?

Only through an astute thoughtful process one can have a clear idea of
what one is attempting to achieve and how to successfully secure the goals of
an extraction process. In researches related to drug discovery of new active
phytoconstituents, extraction is one of the important steps as it is the starting
point for the isolation and purification procedures. An imprudent selection of
an extraction method can actually jeopardize the entire objective of isolating
a bioactive compound and will cause the entire extraction process to fall apart.
Due to often very complex composition of the plant material and the minute
amount of phytoconstituents present, the selection of a pertinent extraction
technique and strategy is of prime importance. Today even though with so
much automation and sophistication being achieved in separation science, it
still can be wasted if an unsuitable extraction strategy is adopted. A poorly
prepared extract is sufficient enough to invalidate even the most powerful
chromatographic technique. Moreover a lot of factors which are involved in
extract production sometimes induce variations in the final product as shown
in Figure 5.2.

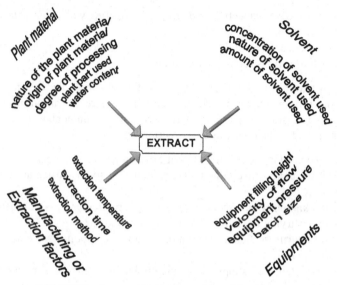

FIGURE 5.2 Factors affecting the overall quality of an extract.

Before going deep into the theories governing extraction process an easy definition for some commonly used terms associated with extraction of botanicals is presented for the convenience of the readers.

Extraction: Extraction so far the term is related to natural product research, involves the separation of medicinally active portions, constituents from a complex biological matrix of either plant or animal origin. The process is generally carried out by the use of suitable solvents of appropriate polarity by using a definite method which may or may not involve the application of heat energy. The products so obtained are relatively impure liquids, semisolids, or powder intended only for oral or external use. Such preparations include infusions, tinctures, decoctions, extracts which are generally known by a common term called galenicals, named after the Greek physician Galen.

Maceration: Soaking a botanical in a suitable solvent(s) for a specific period of time, anywhere from several hours to several days, suitable amount of constituents are dissolved in the menustrum.

Expression: The process of forcibly extracting liquid from solids by the application of mechanical pressure.

Digestion: Maceration with gentle heating at 40–60 °C.

Percolation: A displacement process, whereby a powdered or cut botanical contained in a suitable vessel, having a bottom outlet, is deprived of its soluble constituents by the descent of a solvent through it.

Percolator: A cylindrical or conical vessel with porous diaphragm (wire mesh, gauze, cotton) below, in which the botanical is loaded and its soluble constituents are extracted by the descent of a solvent (menustrum) through it.

Menustrum: A term used to describe the solvent used to extract the botanical of its various constituents (e.g., water, alcohol).

Percolate: The solution coming out from the percolator and containing the extracted substance.

Extractive: The material dissolved from a portion of the botanical when it is solubilized in the menustrum regardless of which of the two extractive processes (water soluble and alcohol soluble) are involved.

Marc: The botanical residue that remains after the extraction, also referred to as the spent herb.

Miscella: It refers to the solution containing the extracted substance.

Solid–liquid extraction: It refers to the extraction process when the powdered botanical is extracted with a solvent.

Liquid–liquid extraction: It refers to the extraction process where two immiscible liquids are involved. The constituents are then separated based upon their partition coefficient between the two immiscible solvents. It is generally performed in a separating funnel.

5.2 UNDERSTANDING THE LINK BETWEEN BOTANICAL EXTRACTION AND THEIR STANDARDIZATION

A key factor in the widespread acceptance of natural or alternative medicine by the international community involves the "modernization" of herbal medicine. So the application of standardization and quality control of herbal materials by use of modern science and technology is a vital issue. Only a standardized and a validated process of extraction alone can assure an extract of consistent quality. In general, standardized extracts are usually tested for a minimum content or range of certain specified marker compounds.

5.2.1 Need for Standardization

- To provide pharmacopoeial standards for pharmaceutically potent raw material \finished products as modern system of medicine is based on sound experimental data, toxicity studies and human clinical studies.
- Current good manufacturing practices (cGMP) for the herbal industry are not well defined nor are the minimum standards of medicinal plant products maintained or regulated.
- The lack of quality standards has resulted in mild to serious adverse effects ranging from hepatotoxicity to death. Hence, herbal ingredients require tools for determining identity, purity and quality and tools have to be technically sufficient, rapid and cost-effective with GMP requirements.
- Quality related problems (lack of consistency, safety, reproducibility and efficacy) seem to be overshadowing the potential genuine health benefits of various herbal products, and a major cause of these problems seems

to be related to the lack of simple and reliable analytical techniques and methodologies for the chemical analysis of herbal materials.

As far as the standardization of extracts is concerned, the aim has to be the reproducibility of all the chemical components contained in an extract, including the unknown ones. An extract normally contains several classes of substances, some of which are active principles, whilst others can be their natural vehicles or they are inert substances. In this context the entire process of extraction and subsequent steps become a crucial step in the analysis of herbs, because it is necessary to extract the desired chemical components (markers) from the herbal materials for further separation and characterization. Chemical standardization often involves chemical identification by spectroscopic or chromatographic fingerprint technique and chemical assay for active constituents or marker compounds if available.

Standardization is a hard task which encompasses a good agricultural practice, good laboratory practices (GLP) for chemical isolation and characterization of active reference substances, validated analytical methods and finally a GMP production of the final ingredient.

5.2.2 New Analytical Techniques for Standardization

- Multicomponent analytical systems like high performance liquid chromatography/electrospray ionization mass spectrometry/nuclear magnetic resonance.
- For evaluation of botanical formulations newer techniques such as DNA fingerprinting, high performance thin layer chromatography, liquid chromatography–mass spectroscopy are widely adopted.
- Real time polymerase chain reaction analysis on a microchip become a standard procedure for the authentication of plant materials.
- Quick, cheap, accurate, and clinically relevant biological systems, mostly micro array-based, demonstrate the level of biological activity for each batch of marketable product.

Thus a key issue in manufacturing herbal products and medicines is standardization. This process requires high knowledge in phytochemical analysis and process technology to ensure the quality assurance required. Thus overall it can be asserted that value addition to herbal extracts and products do increases with an increase in processing and standardization as shown in the Figure 5.3 below.

5.3 GENERAL EXTRACTION APPROACHES AND THEORIES

5.3.1 Classical Approaches for the Extraction of Analytes from Aqueous Sample

5.3.1.1 Liquid–Liquid Extraction

Liquid–liquid extraction (LLE), also known as solvent partitioning technique, is a method to separate compounds based on their relative solubilities in two

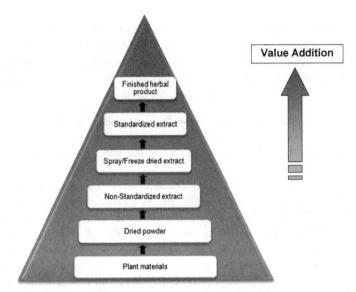

FIGURE 5.3 Increasing value of herbal products with processing and standardization.

different or immiscible liquids, usually water and an organic solvent. Basically, it is an extraction of a substance from one liquid phase into another liquid phase. LLE is the most commonly used approach for the extraction of analytes from aqueous samples. The principle of LLE is that the sample is made to get partitioned between two immiscible solvents in which the sample and matrix have different solubilities. The main advantages of this approach are the wide availability of pure solvents and the use of low-cost apparatus. The technique is frequently used for the separation of an organic product from a reaction mixture after an aqueous work up, or for the isolation of naturally occurring substances. Overall the purpose of LLE is:

1. To separate closed-boiling point mixture (acetic acid, b.p. 118 °C and water, b.p. 100 °C).
2. To separate mixture that cannot withstand high temperature or heat sensitive components (such as antibiotics).

5.3.1.1.1 Theory of LLE

Typical LLE operations utilize the differences in the solubilities of the components of a liquid mixture. The basic steps involved include:

1. Thoroughly mixing of the two immiscible phases.
2. Contacting the feed with the extraction solvent.
3. Allowing selective transfer of solute(s) from one phase to the other.
4. Allowing the two resulting phases to separate out.
5. Recovery of solvent from each phase.

Two terms are used to describe the distribution of an analyte between two immiscible solvents - distribution coefficient and the distribution ratio.

The *distribution coefficient* is an equilibrium constant that describes the distribution of an analyte, A, between two immiscible solvents, e.g. an aqueous and an organic phase. This process can be written as:

$$A \text{ (aqueous layer)} \quad \longleftrightarrow \quad A \text{ (organic layer)}$$

K = Solubility in Organic Layer/ Solubility in Aqueous Layer

The constant K, is the ratio of the concentrations of the solute in the two different solvents once the whole system reaches an equilibrium. At equilibrium the molecules distribute themselves automatically in the solvent where they are more soluble. Inorganic and water soluble materials will stay in the aqueous layer and more organic molecules will remain in the organic layer as shown in Figure 5.4. By choosing a correct solvent system, a molecule can be specifically selected and extracted from another solvent. Using an aqueous and organic (e.g. ethylene acetate, dichloromethane, toluene, chloroform etc.) solvent pairs, the more hydrophobic analytes prefer the organic solvent while the more hydrophilic compounds prefer the aqueous phase.

When this distribution reaches equilibrium, the solute is at concentration $[A]_{aq}$ in the aqueous layer and at concentration $[A]_{org}$ in the organic layer as shown in Eqn (5.1). The distribution ratio of the solute is defined as the ratio of the total analytical concentration of the substance in the organic layer to its total analytical concentration in the aqueous layer, usually measured at equilibrium.

$$D = [A]_{org} / [A]_{aq} \qquad (5.1)$$

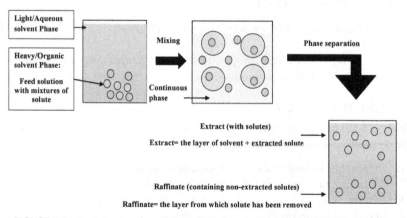

FIGURE 5.4 A schematic representation of solvent extraction (liquid–liquid distribution).

[A]$_{org}$—concentration of A in all chemical forms in organic layer, [A]$_{aq}$—concentration of A in all chemical forms in aqueous form.

5.3.1.1.2 Advantages and Disadvantages of LLE

Advantages of LLE

- Very large capacities are possible with a minimum of energy consumption.
- Heat sensitive products are processed at ambient or moderate temperatures (example: vitamin production).

Disadvantages of LLE

- Large solvent consumption.
- Time/labor intensive.
- May require an evaporation step prior to analysis to remove excess solvent.
- Contamination issues.

5.3.1.1.3 Techniques Involved in LLE

5.3.1.1.3.1 Batch Extraction (Single Stage or Multiple Stage)
Steps:

1. The aqueous feed is mixed with the organic solvent.
2. After equilibration is attained, the extract phase containing the desired solute is separated out for further processing.
3. This can be carried out for example in separating funnel or in an agitated vessel (Figure 5.5).

5.3.1.1.3.2 Continuous Extraction (Cocurrent and Countercurrent) In cocurrent extraction (Figure 5.6), the two phases flow in the same direction between the various contactors.

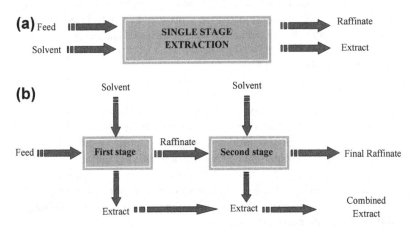

FIGURE 5.5 Schematic representation of (a) single stage, (b) multiple stage batch operation.

FIGURE 5.6 Schematic representation of cocurrent extraction.

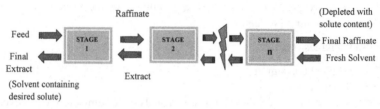

FIGURE 5.7 Schematic representation of countercurrent extraction.

In a countercurrent extraction process (Figure 5.7), the feed to each stage is contacted with solvent from the preceding stage. The feed from Stage 1 is contacted with the extract from Stage 2, and the feed to Stage 2 is contacted with the extract from Stage 3, and so forth. This countercurrent contact results in the gradual enrichment of the solute in the solvent phase across the extraction process.

5.3.2 Classical Approaches for the Extraction of Analytes from Solid Sample

5.3.2.1 Solid–Liquid extraction

Solid–liquid extraction (SLE) is one of the most widely used unit operation techniques in the medicinal and aromatic plant industry. Solid liquid extraction (leaching) means the removal of a constituent from a mixture of solids by bringing the solid material into contact with a liquid solvent that dissolves this particular constituent.

5.3.2.1.1 Mechanism of SLE

SLE involves two steps which are:

- Contacting step: A contact between the solvent and the material to be treated is made, so as to transfer soluble constituents to the solvent. Basically it's a mass transfer which aims at transferring the soluble constituents from the solid phase into the liquid phase by diffusion and dissolution. The solute is first dissolved from the surface of the solid and then subsequently passes into the solution by diffusion. This process may result in the formation of pores in the solid material which exposes fresh (new) surfaces to subsequent solvent penetration to such surfaces.

- Separation step: The solution which is formed from the relatively exhausted solids must be separated. The above two steps may be conducted in separate equipment or in one and the same equipment. Solution resulting from separation step is termed overflow, solids left over are termed underflow.

5.3.2.1.2 Fundamental Aspects of SLE

5.3.2.1.2.1 Concept of Equilibrium Knowledge of both equilibrium and mass transfer concepts is important for efficient and high quantitative recoveries of bioactive compounds from the botanical materials.

$$K = C_e/C_m \tag{5.2}$$

In Eqn (5.2), C_e is the concentration of the bioactive compound in the solvent, C_m is the concentration of the bioactive compound remaining in the dry marc and K is the equilibrium constant, also known as partition coefficient. The value of C_e in turn is dependent on the type of solvent used as well as the temperature, and it dictates the value of K. Thus, the solvent and temperature for extraction are chosen to obtain a high K value for complete or near complete extraction of bioactives from the botanical matrix.

5.3.2.1.2.2 Mass Transfer Concept Mass transfer determines how long the bioactive compound would take to reach the equilibrium concentration in the extracting solvent. During extraction, the concentration of solute inside the solid varies leading to the nonstationary or unsteady condition. There are generally four steps in the extraction of bioactive compounds from botanicals and they are listed below (Figure 5.8).

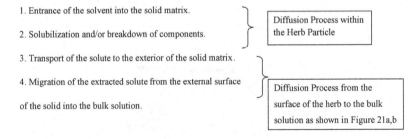

1. Entrance of the solvent into the solid matrix.

2. Solubilization and/or breakdown of components. } Diffusion Process within the Herb Particle

3. Transport of the solute to the exterior of the solid matrix.

4. Migration of the extracted solute from the external surface

of the solid into the bulk solution. } Diffusion Process from the surface of the herb to the bulk solution as shown in Figure 21a,b

- Steady state Mass transfer
  Steady state diffusion refers to a condition when the concentration within the system does not change with respect to time and is described by Fick's first law of diffusion.

$$J_x = -D\, \partial C/\partial x \tag{5.3}$$

- Fick's first law is useful for defining a diffusion coefficient or diffusivity (D).

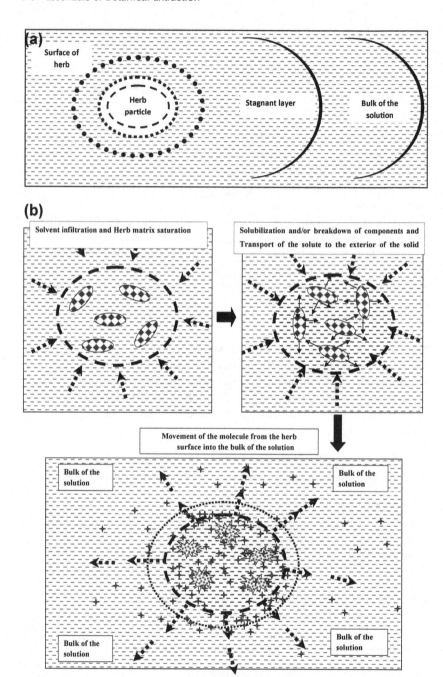

FIGURE 5.8 (a) Diffusion process from the surface of the herb to the bulk solution. (b) Diffusion process from the surface of the herb to the bulk solution.

- It simply establishes that under steady-state conditions (concentration does not change with time) the unidirectional flux of solute (J_x) in the x direction is directly proportional to the diffusivity of the solute, to the area traversed by the flux and to the gradient of solute concentration between two points.
- J is the diffusion flux or mass flux or mass transfer and is directly proportional to the steepness of the concentration gradient.
  - J measures the amount of substance that will flow through a small area during a small time interval.
  - Unit—number of atoms/m^{2}s, moles/cm^{2}s or kg/m^{2}s
- D is the diffusion coefficient which describes the diffusivity of a constituent within the system
  - Unit—[cm^2/s] or [m^2/s].
- C is the concentration (number of particles per unit volume)
  - Unit—atoms/m^3
- x is the direction of flux
- $\partial c/\partial x$ concentration gradient of particles and is the driving force that leads to molecular movement.
- (−) ve sign in this relationship indicates that particle flow occurs in a "down gradient direction," i.e. from regions of higher to regions of lower concentration.

Therefore,

$$J \text{ (moles/m}^2\text{ s)} = -\ D \text{ (m}^2\text{/s)} \ \partial C/\partial x \text{ (atoms/m}^3 \times 1\text{/m)}$$

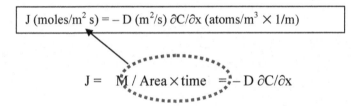

$$J = M \text{/ Area} \times \text{time} = -D\ \partial C/\partial x$$

When differentiated to take the rate of diffusion into account, the flux becomes,

$$J = (1/A)\ (dM/dt) = -\ D\ \partial C/\partial x \tag{5.4}$$

Thus by using the concept of mass transfer, the diffusion rate which is dependent on concentration gradient can be represented as:

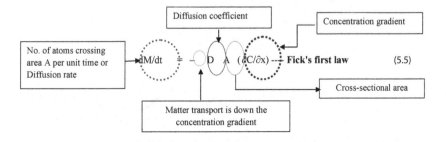

$$\text{No. of atoms crossing area A per unit time or Diffusion rate} \quad dM/dt = -\ D\ A\ (\partial C/\partial x) \quad \text{Fick's first law} \tag{5.5}$$

- Nonsteady state mass transfer model

At the end, the solute/solvent mixture diffuses to the solid surface and finally moves across the stagnant film around the particle to the bulk fluid phase. But in numerous practical cases the concentration and concentration gradient changes with time. Such cases are covered by Fick's second law as shown in Eqn (5.6). Fick's second law predicts how diffusion causes the concentration to change with time.

Fick's second law attempts to express the change in concentration with respect to time with the change in Flux. The rate of change of concentration is dependent on flux and can be represented as:

$$dC/dt = f\,(flux)$$

Consider an elemental volume (box) as shown in Figure 5.9, with a flux of material in and out.

A mass balance on the elemental volume per unit time (both in and out)

$$\Delta mass = \Delta(conc \times vol); \ \Delta flux = \Delta mass/(area\ \Delta time)$$

$$\Delta flux \times area = \Delta mass/\Delta time$$

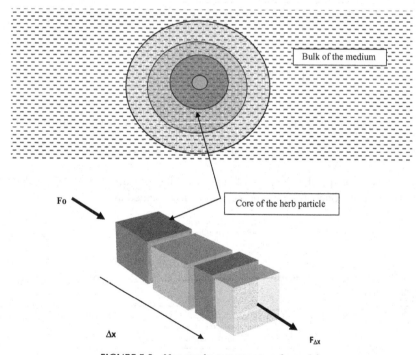

FIGURE 5.9 Non-steady state mass transfer model.

$$\Delta mass = \Delta(conc \times vol); \quad \Delta flux = \Delta mass/(area \; \Delta time)$$

$$\Delta mass/\Delta time = \Delta \; (conc \times vol)/ \; \Delta time$$

$$\frac{d(mass)}{d(time)} = \frac{d(conc * vol)}{d(time)} = -area \bullet \Delta flux$$

$$\Delta V \; dC/dt = -area \bullet \Delta flux$$

$$\Delta V \; dC/dt = -area \bullet \Delta x$$

division by ΔV

$$dC/dt = -\Delta flux/\Delta x$$

as$\Delta x \; -\!\!> zero$

$$\lim_{\Delta x \to o} \frac{1}{\Delta x} \Delta flux = -\frac{dflux}{dx}$$

it appears that

$$dC/dt = - \, dflux/dx$$

$$\frac{\partial C}{\partial t}\bigg|_{x=cons\tan t} = -\frac{\partial F_x}{\partial x}\bigg|_{t=const}$$

$$\frac{\partial C}{\partial t}\bigg|_{x=cons\tan t} = -\frac{\partial F_x}{\partial x}\bigg|_{t=const}$$

$$\frac{\partial C}{\partial t} = -\frac{\partial}{\partial x}\left(-D\frac{\partial C}{\partial x}\right) = D\frac{\partial^2 C}{\partial x^2} \text{------ \textbf{Fick's second law}} \qquad (5.6)$$

5.3.2.1.3 Techniques Involved in SLE

5.3.2.1.3.1 Single Stage Extraction

- Advantage: Ensures a complete contact of the solid feed with the fresh solvent (Figure 5.10).

FIGURE 5.10 Schematic diagram of single stage extraction.

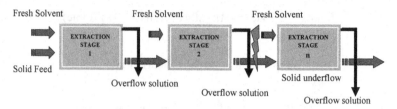

FIGURE 5.11 Schematic diagram of multistage cocurrent extraction.

FIGURE 5.12 Schematic diagram of multistage counter current extraction.

- Disadvantage: Low recoveries of solute obtained and relatively dilute solution produced.
- Troubleshoot:
  - Efficiency of extraction can be improved by dividing the solvent into a number of smaller portions.
  - Carrying out multiple successive extractions instead of only one contact of the entire amount of solvent with the solid.

5.3.2.1.3.2 Multistage Cocurrent Extraction In the multistage cocurrent (parallel) system, fresh solvent and solid feeds are contacted in the first stage. Underflow from the first stage is sent to the second stage, where it comes in contact with more fresh solvent. This scheme is repeated in all succeeding stages (Figure 5.11).

5.3.2.1.3.3 Multistage Counter Current Extraction In the continuous countercurrent multistage system the underflow and overflow streams flow counter-current to each other. This system allows high recovery of solute with a highly concentrated product because the concentrated solution leaves the system after contact with fresh solid (Figure 5.12).

5.4 FACTORS AFFECTING EXTRACTION OF BOTANICALS

5.4.1 Plant Material Related Issue

5.4.1.1 Nature of the Crude Plant Material

The texture or the cellular toughness of the plant material may offer considerable resistance to the entry of solvent during extraction. Plant parts like bark, rhizomes which possess a tougher cellular structure may offer significant amount of resistance to the inflow of extracting solvent. Hence for the extraction of harder/tougher plant parts longer extraction time may be required for the solvent to gain entry inside the cellular channels to dissolve out the constituents. Whereas, the situation may just be the reverse when dealing with softer plant parts such as leaves and stems. Extraction time in this case may be quite lower as the extracting solvent may find an easy entry into the cellular channels of such softer plant parts. Sufficient measures should be taken to keep the plant material free from contamination during the size reduction process.

Fresh or dried plant materials can be used as a source for the extraction of secondary metabolites. Many researchers had reported about plant extract preparation from the fresh plant tissues. The reason behind this came from the ethnomedicinal use of fresh plant materials prevalent among different ethnic groups.

5.4.1.2 Particle Size of the Plant Material

The particle size of the plant material for extraction also needs due consideration. Decreased in particle size will offer enhanced surface area leading to better solute-solvent contact which will facilitate easy entry of the solvent into the plant cellular structure. However, in this case as a common laboratory experience coarse plant material is generally used for extraction. Different kind of mills and cutters are available which should be selected according to the hardness of the plant part to be comminuted.

5.4.1.3 Moisture Content

The moisture content of the plant material depends on the nature of the plant material and also on the drying and postdrying conditions. Considerable amount of moisture can hinder the extraction process. Water can be present as a thin film over the cellular structure which can cause hindrance towards the entry of organic solvents into the cellular structure.

5.4.2 Solvent Characteristics

5.4.2.1 Nature of Solvent

Successful extraction of bioactives from plant material is largely dependent on the type of solvent used in the extraction process. A good solvent must be of less toxic, ease of evaporation at low heat, promotion of rapid physiologic absorption of the extract, inherent preservative potential, inability to cause the extract to complex or dissociate. Solvent choice primarily depends on quantity of phytochemicals to be extracted, what type of targeted compound to be extracted, rate of extraction, diversity of different compounds to be extracted, ease of subsequent handling of the extracts, toxicity of the solvent in the bioassay process, potential health hazard of the extractants etc.

5.4.2.2 Herb–Solvent Ratio

When extraction follows steady-state condition as in the case of maceration, the ratio of herbal drug to the extraction solvent is a decisive factor. The quantity of extractable matter does get increased with an increase in the volume of extraction solvent. Using large volume of extraction solvent results in a slow attainment of steady-state condition.

In case of an exhaustive extraction procedure (percolation), a fixed quantity of drug mass is treated with a variable quantity of solvent until the extractable matter is completely transferred from the herbal drug matrix to the percolate. The ratio of the herbal drug to the extraction solvent may therefore vary from batch to batch within a certain range. It depends on the characteristics of the herbal drug (content of extractable matter, loss on drying, etc.).

5.4.2.3 Solvent under Use

Some of the commonly used organic solvents for herbal extraction are ethanol, methanol, ethyl acetate, diethyl ether, chloroform and petroleum ether. Only two inorganic solvents water and carbon dioxide (as supercritical fluid) are used for herbal extraction. Alcohol or hydro-alcoholic mixture is generally used as an all purpose extracting solvent. However, in herbal drug research it is generally advised to choose a solvent which resembles the form in which the drug is used in crude form among the ethnic groups. In choosing a solvent for extraction, its ability to extract components of a solute has to be considered. For example, ionic solutes can be extracted from aqueous solutions with nonpolar solvents if neutral complexes can be generated in the aqueous phase before extraction. A right choice of solvent with proper selectivity can make the extraction step very efficient and the more efficient the extraction step is, the greater will be the probability for the presence of broader range of compounds present in an extract which later can be separated using different separation and purification techniques. Air or freeze dried samples are normally extracted with a variety of solvents, and sometimes sequentially from low to high polarity, if a crude fractionation of metabolites is sought. It is useful to consider the possibility that the choice of solvent used can determine to some extent whether exocellular or endocellular metabolites or a mixture of both will be extracted. With dried material, ethyl acetate or low polarity solvents will only rinse or leach the sample, whereas alcoholic solvents presumably rupture cell membranes and extract a greater amount of endocellular materials. For instance it was investigated that washing a plant sample with ether afforded a sesquiterpene alcohol and di- and trihydroxyflavones, whereas extraction with methanol provided a different sesquiterpene alcohol and a diterpenesdiol, both of which are also soluble in ether.

Extractions can be either "selective" or "total". The initial choice of the most appropriate solvent is based on its selectivity for the substances to be extracted. In a selective extraction, the plant material is extracted using a solvent of an appropriate polarity following the principle of "like dissolves like". Thus, nonpolar solvents are used to solubilize mostly lipophilic compounds (e.g., alkanes, fatty acids, pigments, waxes, sterols, some terpenoids, alkaloids, and coumarins). Medium polarity solvents are used to extract compounds of intermediate polarity (e.g., some alkaloids, flavonoids), while more polar ones are used for more polar compounds (e.g., flavonoids glycosides, tannins, some alkaloids). In an extraction referred to as "total", a polar organic solvent (e.g., ethanol, methanol, or an aqueous alcoholic mixture) is employed in an attempt to extract as many compounds as possible. This is based on the ability of alcoholic solvents to increase cell wall permeability, facilitating the efficient extraction of large amounts of polar and medium to low polarity constituents. Since the main driving force in herbal extraction is the concentration gradient, hence the herb-solvent ratio plays a key role as mentioned earlier. If insufficient amount of solvent is used for extraction then it may soon get saturated with the constituents resulting in slowing down of the extraction. In such situation the extraction should be restored by adding fresh solvent. Nevertheless, this process will add up to the manufacturing cost particularly when dealing with costly organic solvents. The solubility aspect of the constituents is greatly governed by the polarity of the extracting solvent. Polar solvents will solubilize the polar constituents like alkaloids, glycosides and phenolic compounds. However, sometimes the solubility of less polar substances may be increased with polar solvents, primarily due to the presence of solubilizing compounds like sponins. However, when starting with a plant on whose chemical identity no information is known then successive extraction can be performed starting with the least polar solvent petroleum ether or hexane. Extraction can then be continued using solvents of increasing polarity as per the elutropic series. Several fractions thus collected can then be subjected to phytochemical analysis. The other method is to start the extraction with methanol followed by fractionating with petroleum ether, chloroform, solvent ether and ethyl acetate. These days much is being said about green chemistry. Hence solvent selected should not be detrimental to the environment and also at the same time should not cause biohazards related problems. Long term use of such organic solvents should not be carcinogenic or cause any abnormal change in the normal physiological response of the human body. Boiling point of extracting solvent is very much related to the thermal stability of the constituents. Solvents like dimethyl sulfoxide boil at very high temperature which can be detrimental to the thermolabile phytoconstituents. Solvents like acetonitrile are toxic in nature and hence its presence in the extract even in minute traces would not be desirable.

5.4.3 Manufacturing Related Factors

All these factors are interdependent on each other. An optimized level of all the factors must be determined before starting of the extraction process to ensure maximum yield. However, knowledge regarding the chemical nature and thermal stability of the constituents is very essential in order to determine the optimum blend of the operating conditions.

FURTHER READING

Kahol, A.P., Tandon, S., Singh, K.L., 1998. Developments in separation technologies for the processing of medicinal and aromatic plants. Perfumes Flavours Assoc. India J. 20 (3), 17–28.

Chapter 6

Classification of Extraction Methods

Chapter Outline

Essentials of Botanical Extraction. http://dx.doi.org/10.1016/B978-0-12-802325-9.00006-9

6.1 CLASSIFICATION OF VARIOUS NONCONVENTIONAL EXTRACTION TECHNIQUES

6.1.1 Analytical Extraction Methods

It refers to a complete exhaustive extraction from which quantitative results can be obtained.

6.1.1.1 Microwave Assisted Extraction

Microwaves heat up the molecules of any object by dual mechanism of ionic conduction and dipole rotation. Microwaves are the nonionizing electromagnetic waves positioned between the X-ray and infrared rays in the electromagnetic spectrum with frequency between 300 MHz and 300 GHz as depicted in Figure 6.1 below. The two types of oscillating perpendicular fields that generate microwaves are the electric field (e) and magnetic field (H).

When the microwaves interact with polar solvents, heating of substance is caused due to any one of the above mentioned phenomena, individually or simultaneously. The migration of ions under the influence of the changing electric field is called Ionic conduction. If the solution offers a resistance to this migration of ions, a friction is generated and the solution gets heated up. The realignment of the dipoles of the molecule with the rapidly changing electric field is called Dipole rotation. Whenever dipolar molecules are subjected to a microwave field, they join electrostatically and tend to align themselves with the field mechanically. Since the microwave field is alternate in direction at a certain frequency, the dipoles will attempt to realign as the field reverses, and so are in a constant state of mechanical oscillation at

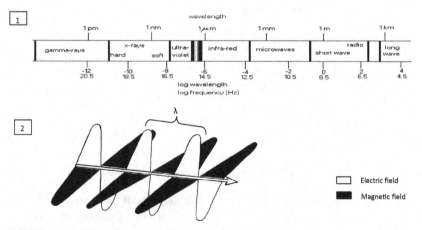

FIGURE 6.1 Position of Microwave in an Electromagnetic spectrum (a) and components of a microwave (b) Lambda (λ), wave length.

the microwave frequency as shown in the Figure 6.2 where the interaction between water molecule (a dipole) and microwave has been shown.

At 2.45 GHz the dipoles align and randomize at 5×10^9 times/sec and this forced molecular movement causes heat, due to friction. No heating occurs when the frequency is greater than or less than 2450 MHz as the electrical component changes at a much higher or lower speed. The noteworthy observation from the above mentioned mechanism is that only dielectric material or solvents with permanent dipoles get heated up under microwave. The value of dissipation factor (tan β), is a measure of the efficiency with which different solvents heat up under microwave. The dissipation factor is given by the Equation (1.1)

$$\tan \beta = \varepsilon'' / \varepsilon \qquad (1.1)$$

Where, ε'' denotes efficiency with which microwave energy is converted to heat i.e., the dielectric loss, ε is the measure of the ability to absorb microwave energy known as dielectric constant. Physical parameters of some solvents commonly used in microwave extraction are given in the following Table 6.1.

Extraction Principle: Dried plant material is generally used for extraction in most cases, but still plant cells contain small traces of moisture that serves as the target for microwave heating. Owing to microwave effect when moisture gets heated up inside the plant cell, it evaporates and generates tremendous pressure on the cell wall from within. The pressure pushes the cell wall from inside, stretching and ultimately rupturing it, which helps in leaching out of the

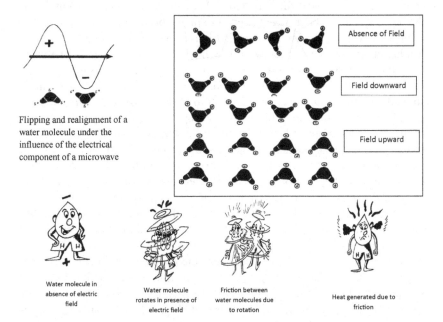

FIGURE 6.2   Interaction between water molecule and microwave.

TABLE 6.1 Physical Parameters of Some Solvents Commonly Used in Microwave Extraction

Sr. no	Solvent	Dipole moment at 20 °C	Dielectric constant at 20 °C	Dissipation factor at 2.45 GHz
1	Acetonitrile	3.44	37.5	–
2	Ethyl acetate	1.88	6.02	–
3	Water	1.84	80.4	0.123
4	Methanol	1.70	33.7	0.659
5	Ethanol	1.69	25.7	0.941
6	1-Butanol	1.66	–	0.571
7	Acetone	–	21.4	–
8	Diethyl ether	–	4.389	–
9	Chloroform	–	4.8	–
10	Hexane	<0.1	1.88	–

active constituents from the ruptured cells to the surrounding solvent and in the process helps in improving the yield of phytoconstituents. This process can even be more intensified if the plant matrix is soaked with solvents with higher heating efficiency under microwave (i.e. higher tan δ value). Higher temperature attained during this entire process can hydrolyze ether linkages of cellulose which is present in the cell wall and in the process reduces its mechanical strength and this in turn helps the solvent to have an easy access for the compounds present inside the cell. During the rupture process, a quick exudation of the chemical compounds from within the cell into the surrounding solvents takes place. This mechanism of microwave extraction based on exposing the analytes to the solvent through cell rupture is different from that of heat-reflux extraction which depends on a series of permeation and solubilization processes to bring the analytes out of the plant matrix. Moreover, the transport of dissolved ions increases solvent penetration into the matrix, thus helping in the release of phytochemicals. Research has shown that during the extraction of essential oils from plant materials, microwave assisted extraction (MAE) allows desorption of compounds of interest out of the plant matrix. This is due to the targeted heating of the free water molecules present in the gland and vascular systems; which leads to localized heating causing remarkable expansion, with subsequent cell wall damage, allowing essential oil to flow towards the organic solvent. The consequence of microwave heating is strongly dependent on the dielectric susceptibility of both the solvent and solid plant matrix. Mostly the sample is placed in a single solvent or mixture of solvents that absorb microwave energy strongly. Temperature helps in increasing the penetration power of the solvent into the matrix and as a result constituents are released into the surrounding hot solvent. However, in some cases only selective heating of sample matrix is brought about by immersing the sample in a microwave transparent solvent (hexane, chloroform). This approach is particularly useful for thermolabile components to prevent their degradation.

Instrumentation: There are two types of commercially available MAE systems: closed extraction vessels and focused microwave ovens. The close type performs extraction under controlled pressure and temperature. The latter is also named as focused microwave assisted Soxhlet or solvent extraction (FMASE), in which only a part of the extraction vessel containing the sample is charged with microwave (Figure 6.3). However, both the above-mentioned systems are available as multimode and single mode or focused systems. A multimode system allows random scattering of microwave radiation within the microwave cavity, so that every corner in the cavity and sample it contains receives even irradiation. Single mode or focused systems allow focused microwave radiation on a particular zone where the sample is subjected to a much stronger electric field than in the previous case (Figure 6.4). Even a modified multimode domestic microwave oven operates as an open vessel extraction system (Figure 6.5).

Principle elements of a microwave device are listed below. Both multi mode and focused microwave devices comprise of four major components:

1. Microwave generator: magnetron generates microwave energy.
2. Wave guide to propagate the microwave from the source to the microwave cavity.
3. Applicator to place the sample.
4. Circulator allows the microwave to move only in the forward direction.

The applicator in case of multimode system can be a closed cavity inside which microwaves are randomly dispersed. Even distribution of microwave

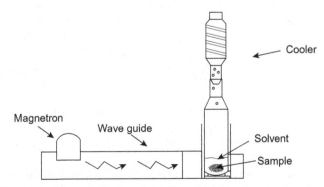

FIGURE 6.3 Schematic diagram of open focused microwave system for extraction.

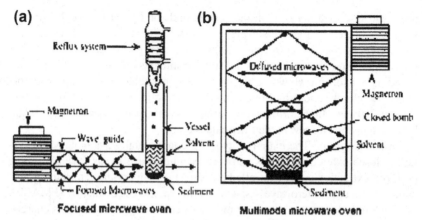

FIGURE 6.4   Monomode (a) versus multimode systems (b).

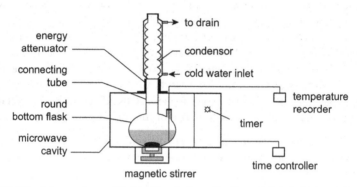

FIGURE 6.5   Scheme of a modified multimode domestic microwave oven (open vessel extraction).

energy inside the cavity can be achieved by using beam reflectors that makes sample heating independent of the position. In focused microwave systems, the extraction vessel is kept directly in a microwave waveguide which at the same time also acts as an applicator. The bottom of the vessel is directly exposed to the microwaves, whereas the upper region of the vessel remains cool as glass is transparent to microwaves and hence does not get heated up in the process. The advantages of closed-vessel systems are as follows.

1. Can reach higher temperatures than open vessel systems because the increased pressure inside the vessel raises the boiling point of the solvents used. The higher temperatures actually decrease the time needed for the microwave treatment.
2. Loss of volatile components during microwave treatment can be avoided.
3. Requires less solvent as no evaporation occurs.

4. Contamination can be avoided by preventing the airborne particles to enter the system.
5. The fumes produced during an acid microwave extraction are contained within the vessel, so no extra provision for handling potentially hazardous fumes need to be made.

However, closed-vessel systems are subject to several drawbacks which are as follows.

1. Explosion risks due to high pressure generated inside the vessel.
2. A limited amount of sample only can be processed.
3. Material of the vessel which is polytetraflouro ethylene, does not allow high solution temperatures.
4. Due to its single-step procedure, the addition of reagents or solvents during operation cannot be performed.
5. The vessel must be cooled down before it can be opened after the treatment to avoid loss of volatile constituents.

6.1.1.1.1 Factors Affecting MAE

6.1.1.1.1.1 Solvent Nature and Volume Solvent choice for MAE depends upon the solubility of the analyte under study, interaction between solvent and plant matrix and microwave absorbing properties of the solvent.

Suitably, the solvent should have a high selectivity towards the analyte excluding unwanted components. Nonpolar solvent are generally transparent to microwave and so does not heat up under microwave irradiation, whereas polar solvents have good microwave absorbing capacity and hence heats up faster and can speed up the extraction process. Notably dielectric properties of the solvent towards microwave heating play a pivotal role in microwave extraction. It was recognized that both the efficacy and selectivity of MAE greatly depend on the dielectric constant of the extracting solvent or its mixture. Generally, in most of the cases mixtures of solvents with good heating efficiency under microwave (high $\tan\delta$ value) are used and aqueous methanol and ethanol serves the purpose to its best. Inner glandular and vascular network of the plant material are highly susceptible to microwave energy due to their high natural moisture content. Instant internal heating of these structures brings about effective cell rupture, releasing the analytes into the solvent. Solvent free MAE has been designed for the extraction of volatile oil from several aromatic herbs where the natural moisture content of the plant material serves as the heating source and no extracting solvent are used.

Volume of the solvent is also a crucial factor. The overall idea is that the solvent volume must be sufficient enough to ensure that the plant matrix is always entirely immersed in the solvent throughout the entire extraction time. There are many reports regarding the volume of solvent to be used with respect to the amount of sample. In most of the cases, a higher ratio of solvent volume to solid matrix may be effective in conventional extraction methods. However, in MAE a higher ratio may yield lower sample recoveries, which may be due to inadequate stirring of the solvent by microwaves. The amount of plant materials and the volume of extraction solvent used in the MAE reported before were generally ranged between a laboratory scale of milligram and milliliter. The heating efficiency of the solvent should also be given due importance as the evaporation of the solvent will depend how rapidly it heats up under microwave. Therefore a careful optimization of this parameter is of primary importance in MAE.

Extraction time: Like other extraction technique, time is another factor whose influence needs to be taken into account. Generally by increasing the time of extraction, the quantity of analytes extracted may get increased, but there exists a risk of degradation. Often 15–20 min is sufficient, but even 40 s have been demonstrated to have given excellent sample recovery. An appropriate study on optimization of extraction time is vital because extraction time may vary with different plant part used and with the type of extraction technique used. Irradiation time is also influenced by the dielectric properties of the solvent. Solvents like ethanol, methanol and water may heat up extremely on longer exposure thus risking the thermolabile components.

Microwave Power: Microwave power and extraction time are two such factors, which influences each other to a great degree. A blend of low or moderate power with longer exposure may be a wise approach. Rapid breakdown of cell wall takes place at higher temperature when kept at higher power, as a result together with the desired analytes impurities are also leached out into the solvent. Whereas at low power levels, the cell wall rupture might take place gradually and enables selective MAE. In a closed vessel system, the chosen power settings depends on the number of samples to be extracted during a single extraction run, as up to 12 vessels can be treated in a single run. The power must be chosen sensibly to avoid excessive temperature, which could lead to sample degradation and overpressure inside the vessel causing explosion threat.

Matrix characteristics: The particle size of the sample and the state in which it is presented for MAE can have a colossal effect on the recoveries of the compounds. Fine powders can augment the extraction process by providing larger surface area, which provides better contact with the solvent, also finer particles will allow improved or much deeper penetration of the microwave energy. One of the difficulties associated with the use of finer particles is the problem of separation of the analyte from the solvent after microwave treatment. In most of the cases, centrifugation or filtration is applied to eradicate the above problem. It has been reported that sample pretreatment prior to MAE can bring about effective and selective heating of the plant matrix. The sample may be selectively heated by microwave with the extracting solvent surrounding the sample being transparent to microwave. Such approach can be effectively used for thermolabile materials. In many cases the natural moisture content of the sample matrix improves the extraction efficiency, as in the case of extraction of essential oil. In some cases soaking of the dried plant material in the extracting solvent prior to MAE has resulted in improved yield. This phenomenon is called preleaching extraction.

Temperature: Microwave power and temperature are very much related to each other and needs special attention particularly when working with closed vessel system. In a closed vessel system, the temperature may rise well above the boiling point of the solvent under use. This increased temperature indeed results in improved extraction efficiencies as desorption of sample from active sites in the matrix will increase. Moreover, solvents have higher capacity to solubilize analytes at higher temperature condition while surface tension and solvent viscosity decreases simultaneously, which will improve sample wetting and matrix penetration respectively.

6.1.1.1.2 Applications and Comparison with Conventional Techniques

Application of MAE for open and closed vessel systems have been tabulated separately in the following Table 6.2.

A comparison between microwave assisted extraction and conventional extraction has demonstrated in Table 6.3which suggests the superiority of the former in extraction of botanicals.

Poor extraction yield due to thermal degradation and oxidation of some active compounds while performing conventional microwave extraction has led to the development of more efficient MAE. These modifications are discussed hereinafter in the following sections.

- Nitrogen-protected microwave-assisted extraction (NPMAE)
  Oxidation of the active compounds during the extraction process can be prevented by using a pressurized inert gas, such as nitrogen and argon, in a closed system. Hence, oxidizable compounds under the inert condition results in a higher extraction yield. For instance, nitrogen-protected microwave assisted extraction (NPMAE) uses nitrogen to pressurize the extraction vessel. This technique has been employed in the extraction of ascorbic acid from guava, yellow pepper, green pepper and cayenne pepper. Highest extraction yield was reported in NPMAE as compared to typical MAE and Soxhlet extraction due to the protection effect exerted by nitrogen which prevents oxidation of the active compounds.
- Vacuum microwave-assisted extraction
  Extraction of thermal sensitive compounds using mild operating conditions i.e. low pressure and temperature can be carried out in vacuum microwave-assisted extraction. This type of MAE enhances mass transfer mechanism by promoting diffusion of active compounds to the solvent via the suction pressure. The risks of thermal degradation and oxidation of the active compounds can be minimized by introducing vacuum pressure, as it lowers the associated boiling temperature of the solvent.
- Ultrasonic microwave-assisted extraction
  Enhancement of mass transfer mechanism in extraction can be achieved by another type of MAE known as ultrasonic microwave assisted extraction (UMAE). Additional ultrasonic wave emitted by UMAE intensifies mass

TABLE 6.2 Application of MAE for Open and Closed Vessel Systems

Plant material	Analyte	Solvent	Extraction time	Country
Open vessel MAE performed with domestic (modified/unmodified) microwave set up				
Cuminum cyminum and Zanthoxylum bungeanum	Essential oil	Solvent free extraction	30 min	China
Whole plant of Nothapodytes foetida	Camptothecin	90% Methanol	7 min	India
Fresh stems and leaves of Lippia alba	Essential oil	Solvent free extraction	30 min	Colombia
Dry fruits of Macleaya cordata	Sanguinarine and chelerythrine	0.1 molelit^{-1} HCl Aqueous solution	5 min	China
Dried roots of Salvia miltiorrhiza	Diterpenes like tanshinones	95% Ethanol	2 min	China
Dried bark of Eucommia ulmodies	Geniposidic acid and chlorogenic acid	Methanol water mixture	40 s	China
Tobacco leaves	Solanesol	Hexane:ethanol (1:3)	40 min	China
Licorice root	Glycyrrhizic acid	50–60% Ethanol with 1–2% ammonia	4–5 min	China
Dried apple pomance	Pectin	HCl solution	20.8 min	China
Dried berries of Embeliaribes	Embelin	Acetone	80 s	India
Curcuma rhizomes	Curcumin	Acetone	4 min	India
Green tea leaves	Polephenols and caffeine	50% Ethanol water mixture	4 min	China
Artemisia annua L.	Artemisnin	#6 Extraction oil	12 min	China

Continued

TABLE 6.2 Application of MAE for Open and Closed Vessel Systems—Cont'd

Plant material	Analyte	Solvent	Extraction time	Country
Closed vessel MAE performed on different plant materials				
Pastinaca sativa fruits	Furano coumarins	80% Methanol	31 min	Poland
Flowering tops of *Melilotus officinalis* L.	Coumarin and melilotic acid	50% Aqueous ethanol	10 min	Italy
Hypericumper foratum and *Thymus vulgaris*	Phenolic compounds	Aqueous HCl	30 min	Czech republic
Capsicum annum powders	Pigments	Acetone: water (1:1)	120 s	Portugal
Dried ginger roots	Ginger	Ethanol water mixture	60–120 s	Venezuela
Roots of *Panax ginseng*	Saponin	60% Ethanol	30 s	South Korea and Canada
Dry needles of *Taxus baccata*	Paclitaxel	90% Methanol	15 min	Iran
Leaves of *Oleaeuropeaea*	Oleuropein and related biophenols	Ethanol:water (80:20)	8 min	Spain
Cicerarietinum L. (chickpea seeds)	Saponin	Ethanol	20 min	Israel
Panax ginseng roots	Ginsenosides	Ethanol	15 min	Taiwan
Roots of *Morinda citrifolia*	Anthraquinones	80% Ethanol in water	15 min	Thailand
Olive seeds	Edible oil	Hexane	20–25 min	Spain
Capsicum frutescens	Capsaicinoids	Ethanol	5 min	Spain

TABLE 6.3 A Comparison Between Microwave Assisted Extraction and Conventional Extraction

Sr. No	Conventional Methods of Extraction	Microwave Assisted Extraction
1	Heating proceeds from the vessel surface into the inside of the reaction vessels.	Since reaction vessel is transparent to microwave so the contents of the vessel are heated simultaneously resulting in volumetric heating.
2	The vessel must be in physical contact with the heat source.	No need of physical contact of the reaction vessel with the higher temperature source.
3	Heating takes place by thermal source.	Heating takes place by electromagnetic wave.
4	Heating mechanism is conduction followed by a convection current inside the vessel.	Heating mechanism is dielectric polarization and ionic conduction.
5	Transfers of energy occur from the surface of vessel ⟶ to the mixture ⟶ to reacting species.	The core mixture is heated directly while surface acts as the medium of heat loss.
6	All components of the reaction mixture receives same amount of heat energy.	In microwave heating specific components of the reaction mixture can be heated to different extents depending upon their capacity to absorb microwave.
7	Heating rate is less.	Heating rate is several folds high.
8	In conventional heating, the highest temperature that can be achieved is controlled by the boiling point of a particular mixture.	In microwave, the temperature of the mixture can be raised more than its boiling point and thus it is independent of solvent's boiling point.

transfer mechanism as the combined microwave and ultrasonic waves provides high momentum and energy to rupture the plant cell and leach out the analytes into the extraction solvent. As a result, extraction proceeds with shorter extraction time and lower solvent consumption. UMAE has been used to extract a variety of active compounds such as lycopene from tomatoes, vegetable oil and polysaccharides from various plants.

- Dynamic microwave-assisted extraction

  All the methods discussed so far have separate extraction step and analytical step. Both the steps work independently and required high expertise of the operator to collect and clean up the extract prior to analysis. The clean up procedure is time-consuming as it involves several steps to concentrate the extract. This can be improved by modifying the extraction process in a continuous and automatic manner and coupling online with analytical step. With that, dynamic microwave-assisted extraction (DMAE) has been developed where both the extraction and analytical steps can be carried out in a single step.

DMAE offers fast extraction and lower solvent consumption over conventional techniques such as reflux extraction and ultrasonic extraction. Due to the fluidized state of extraction solvent-sample system, DMAE promotes rapid transfer of microwave energy to the extraction solvent and the sample. The need of extraction cycle is eliminated and replaced by continuous extraction. Thus, the overall extraction time is reduced. Besides, the risks of analyte loss and contamination can also be minimized as the system runs continuously in a closed and automated manner.

- Solvent-free microwave-assisted extraction

  In view of other modified MAEs, solvent-free microwave assisted extraction (SFME) is commonly used for essential oil extraction and water can be incorporated to extract targeted compounds or even in case of fresh plant material the high natural moisture content can be made use of. SFME significantly reduces extraction time as compared to conventional methods from few hours to 20–30 min for essential oil extraction. In many cases, the quality of essential oil obtained by conventional methods is affected by oxidation and hydroxylation of active compounds due to long extraction time and high water content. Therefore, essential oil extracted by SFME is considered a better choice. For instances, the essential oils extracted by SFME from basil, garden mint and thyme are more valuable compared to those extracted by hydro distillation because higher amount of oxygenated compounds are present.

Microwave assisted extraction encompasses following advantages, over the conventional heating.

1. Uniform heating occurs throughout the material as opposed to surface and conventional heating process.
2. Desirable chemical and physical effects are produced.
3. Floor space requirement is nominal.
4. Better and more rapid process control can be achieved.
5. Selective heating in MAE increases efficiency and decrease operating cost.
6. High efficiency in heating can be achieved.
7. Unwanted side reaction can be avoided.
8. No loss of heat to the environmental.

The major disadvantages of microwave assisted extraction are:

1. Microwave extraction technique is limited in terms of solvents and nature of the solid material. The solvent must possess dielectric property and must absorb microwaves to give better yield.
2. The efficiency of MAE is typically low when the solvent lacks a significant dipole moment for microwave energy absorption.

3. Thermolabile samples cannot be extracted by directly applying microwaves as degradation of solute has been observed.

6.1.1.2 Ultrasound Assisted Extraction

Over the past two decades, applications of ultrasound in chemistry as well as in food and pharmaceutical industries have become an exciting new field of research. Usually, the traditional techniques require long extraction time and have low efficiency. Moreover, many natural products are thermolabile and may degrade during thermal extraction. The use of ultrasound resulted in increased extraction in a shorter time and at lower temperature conditions. The mechanical effect of ultrasound is its capability to speed up the extraction of active plant compounds, contained within the body of matrix, due to disruption of the cell walls and thus enhances mass transfer. Recently ultrasound assisted extraction (UAE) has been employed to extract pharmaceutically active compounds, polysaccharides, saturated hydrocarbons, cellulose, flavonoids, fatty acid esters and steroids from plant materials. The isolation and purification of compounds from crude plant extracts through classical methods are both costly and time-consuming.

UAE is a process that uses acoustic energy (a mechanical energy i.e. it is not absorbed by molecules but is being transmitted throughout the medium) and solvents to extract target compounds from various plant matrices. Ultrasound is transmitted through a medium via pressure waves by inducing vibrational motion of the molecules which alternately compress and stretch the molecular structure of the medium due to a time-varying pressure. The sound waves that propagate into the liquid media result in the production of an alternating high-pressure (compression) and low-pressure (rarefaction) cycles, whose rates depends on the frequency of vibration of sound wave. Compression cycles (due to high pressure) push molecules together, whereas expansion cycles (due to low pressure) pull them apart. In a liquid, the expansion cycle produces negative pressure that forces molecules away from one another. If the ultrasound intensity is high enough, the expansion cycle can create bubbles or cavities in the liquid. The process by which bubbles form, grow and undergo implosive collapse is known as "cavitation". The steps involved in the process are depicted in Figure 6.6.

During the implosion very high temperatures (approx. 5000K) and pressures (approx. 2000atm) are reached locally. The implosion of the cavitation bubble also results in sharp liquid kinetic jets of up to 280 m/s velocity. The resulting shear force breaks the cell envelope of the sample present within the medium mechanically and improves material transfer.

6.1.1.2.1 Factors Affecting Cavitation

The frequency of ultrasound is an important parameter and influences the bubble size. At lower frequencies (20 kHz), the bubbles produced are larger in size and when they collapse higher energies are produced. At higher frequencies, bubble formation becomes more difficult and cavitation does not occur. As a rule, increasing the intensity increases the sonochemical effects. Thus cavitation bubbles, which

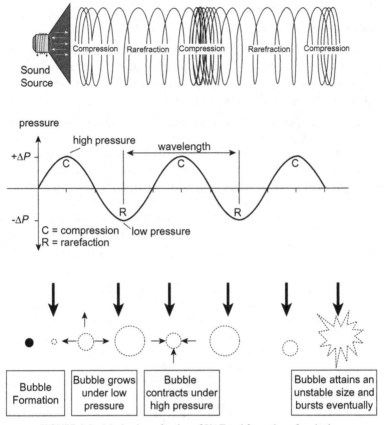

FIGURE 6.6 Mechanism of action of UAE and formation of cavitation.

were initially difficult to create at the higher frequencies as a result of the shorter duration of rarefaction cycles, are now possible. On the other hand temperature of the treatment medium, viscosity and frequency of ultrasound determines the intensity of bubble collapse. As temperature increases, cavitation bubbles develop more rapidly, but the intensity of collapse gets reduced. This is thought to be due to an increase in the vapor pressure, which is offset by a decrease in the tensile strength. This results in cavitation becoming less intense and therefore less effective as temperature increases. Therefore, in order to maximize sonochemical effects any experiment should be conducted at as low a temperature as possible or by using a solvent of low vapor pressure.

This effect can be overcome if required, by the application of an overpressure (200–600 kPa) to the treatment system. Combining pressure with ultrasound and heat increases the amplitude of the ultrasonic wave and it has been shown that this can increase the effectiveness of microbial inactivation in food processing industries.

6.1.1.2.2 Instrumentation

An ultrasonic extractor will always consist of the following parts:

1. **A generator:** This is an electronic or mechanical oscillator that needs to be rugged, robust, reliable and able to operate with and without load.
2. **A transducer:** this is a device for converting mechanical or electrical energy into sound energy at ultrasonic frequencies.

Following are the two different types of ultrasound equipments which are commonly used in laboratory.

The first one is the ultrasonic cleaning bath (Figure 6.7) which is commonly used for solid dispersion into solvent and for degassing solutions. The second one, the ultrasonic probe or horn system (Figure 6.8), is much more powerful because the ultrasonic intensity is delivered on a small surface (only the tip of the probe) compared to the ultrasonic bath. This system of probe is widely used for sonication of small volumes of sample but special care has to be taken because of the fast rise of the temperature into the sample occurs.

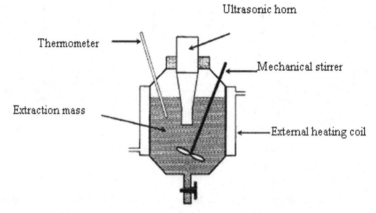

FIGURE 6.7 Direct sonication using an ultrasonic horn (probe type).

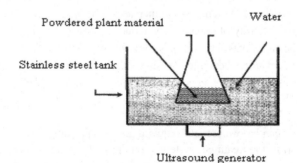

FIGURE 6.8 Indirect sonication using an ultrasonic bath.

6.1.1.2.2.1 Advantages and Disadvantages of Ultrasonic Baths are as Follows Although the cleaning bath is the piece of ultrasonic equipment most widely used by chemists, it is not necessarily the most effective. The advantages of using an ultrasonic bath are as follows:

1. The ultrasonic bath is the most widely available laboratory source of ultrasonic radiation.
2. Small cleaning baths are inexpensive.
3. The acoustic field is evenly distributed throughout the bath liquid.
4. No special adaptation of chemical apparatus is required. This allows conventional glassware to be used and facilitates the addition of chemicals, the use of high or low pressures or even an inert atmosphere during the operation.

On the other hand, the disadvantages of using an ultrasonic bath can be summarized as follows:

1. The amount of power dissipated from the bath into the analytical system is usually not very large.
2. The energy input must be assessed on an individual basis for each system as the amount of power actually delivered will depend on the bath size, the vessel type in batch steps or manifold type in continuous steps, wall thickness and bath position.

6.1.1.2.2.2 Advantages and Disadvantages of Ultrasonic Probes are as Follows Probe devices undoubtedly provide the most efficient method for transmitting ultrasonic energy into an analytical process or step.

The advantages of using an ultrasound probe for this purpose are as follows:

1. The ultrasonic power delivered by a horn is directly related to the magnitude of vibration of the tip. Maximum powers of several 100 Watt per square centimeter can thus be easily achieved.
2. Ultrasonic streaming from the tip of the probe operated at moderate power is often sufficient to provide bulk mixing when dipped in the target system since energy losses during the transfer of ultrasound through the bath media and reaction vessel walls are eliminated.
3. The probe can be tuned to give optimum performance (tuning here is the process whereby the entire probe assembly is brought into resonance with the transducer).

On the other hand, disadvantages of ultrasound probes in this context include the following:

1. Metal particles eroded from the tip can contaminate the system.
2. The high intensity of irradiation in the zone close to the tip may produce radical species potentially interfering with the normal course of the experiment.

6.1.1.2.3 Advantages and Disadvantages of UAE

Ultrasound-assisted extraction is a simple, inexpensive and efficient alternative to conventional extraction techniques. The main benefits of use of ultrasound in solid–liquid extraction include the faster kinetics and increase of extraction yield. Ultrasound can also reduce the operating temperature allowing the extraction of thermolabile compounds. Compared with other novel extraction techniques such as microwave assisted extraction, the ultrasound apparatus is cheaper and is quite ease in operation. Moreover ultrasound-assisted extraction, like Soxhlet extraction, can be used with any solvent for extracting a wide variety of bioactives. However, the effects of ultrasound on extraction yield and extraction kinetics may be linked to the nature of the plant matrix. Therefore, those two factors must be considered carefully in the design of ultrasound assisted extractors. We hereby present a detailed comparison chart of UAE with other well known extraction processes.

Methods based on Soxhlet extraction have traditionally been used as references to assess the performance of methods based on other principles, including ultrasound. The main advantages of UAE over conventional Soxhlet extraction are as follows.

1. Cavitation increases the polarity of the system, including extractants, analytes and matrix; this increases the extraction efficiency, which can be similar to or greater than that of conventional Soxhlet extraction.
2. Ultrasound-assisted leaching allows the addition of a coextractant to increase further the polarity of the liquid phase.
3. It also allows the leaching of thermolabile analytes, which are altered under the working conditions of Soxhlet extraction.
4. The operating time is invariably shorter than with Soxhlet extraction.

However, ultrasound-assisted leaching is at a disadvantage with respect to Soxhlet extraction in other respects, as follows.

1. In batch systems, which are the most widely used, the solvent cannot be renewed during the process, so its efficiency is a function of its partitioning coefficient.
2. The need for altering and rinsing after extraction lengthens the overall duration of the process, and increases solvent consumption and the risk of losses and/or contamination of the extract during handling.

UAE provides several interesting advantages over microwave-assisted leaching, as follows:

1. It is sometimes faster.
2. In acid digestions, the ultrasonic procedure is safer than the microwave one, as the former does not require high pressure or high temperature.
3. In many cases, the ultrasonic procedure is simpler, as it involves fewer operations and is thus less prone to contamination.

However, ultrasound-assisted extraction has the following shortcomings when compared to microwave assisted extraction.

1. It is usually less robust, as aging of the surface of the ultrasonic probe can alter extraction efficiency.
2. Particle size is a critical factor in ultrasound assisted applications.

6.1.1.3 Supercritical Fluid Extraction

During recent years, there has been a surge in the use of supercritical fluids as an extraction medium in the field of environmental science and natural product chemistry. The existing extraction techniques by liquid solvents are of environmental concern and supercritical fluid extraction (SFE) technology is providing reliable, safe, selective and above all, cost-effective technique over that of other conventional extraction techniques. Today, supercritical fluid extraction using carbon dioxide as the supercritical fluid is a well established method for the extraction of botanicals.

6.1.1.3.1 Theory

SFE is an advanced separating technique based on the improved solvating power of certain gasses above their critical point. Above its own critical temperature and pressure, the substance becomes a supercritical fluid and possesses the properties of gas as well as that of a liquid. Thus the gas-like mass transfer and the liquid like solvating power of a supercritical fluid give them an edge over other extracting solvents.

Supercritical fluids are produced by heating a gas above its critical temperature or compressing a liquid above its critical pressure. So a pure supercritical fluid is any compound at a temperature and pressure above the critical point. Above the critical temperature of a compound, the pure gaseous component cannot be liquefied regardless of the pressure applied. The minimum pressure required to liquefy a substance at its critical temperature is the "critical pressure". The critical temperature and pressure combine to define a unique point on the phase diagram known as the critical point (Figure 6.9).

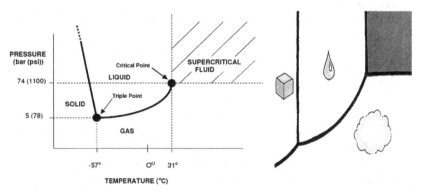

FIGURE 6.9 Schematic depiction of phase diagram.

Super critical fluid shows a pressure-tunable dissolving power. They have a liquid like density, and is having a gas like transport properties. Less viscous nature helps supercritical fluid to diffuse more quickly than liquids into the botanical material and can therefore extract plant products more effectively and faster than conventional organic solvents. By increasing the pressure of the gas above the critical point, it is quite possible to produce liquid like densities and solvating strengths. With an increase in pressure near the critical point, the density of the gas also rises rapidly. Under such condition, the solubility of numerous compounds is several orders of magnitude greater than predicted from the classical thermodynamics of ideal gasses. As the mean distance between molecules decreases, a nonideal gas behavior will begin to control the interactions between the solvent and the compound accounting for incredible increase in solubility. In supercritical region, solvating strength is a direct function of density, which in turn is dependent on the pressure of the system. So the rule is higher the density greater is the solvent strength. Supercritical fluid extraction occurs in a manner similar to liquid extraction except the fact that the solvent is a supercritical fluid. Supercritical fluid is also referred to as dense gasses.

Some major advantages of supercritical fluid extraction are as follows:

1. Supercritical fluids have relatively lower viscosity and higher diffusivity. Therefore, it can penetrate into porous solid materials more effectively than liquid solvents and, consequently, it may render much faster mass transfer resulting in faster extractions. For example, with comparable or better recoveries, the extraction time could be reduced from hours or days in a solid–liquid extraction to a few minutes in SFE.
2. SFE can provide quantitative or complete extraction as fresh fluid is continuously forced to flow through the samples.
3. In SFE the solvation power of the fluid can be altered by changing pressure and/or temperature to achieve high selectivity. This solvation power of supercritical fluids is particularly useful for the extraction of complex samples such as plant materials.
4. SFE can eliminate the additional sample concentration step thus preventing time and organic solvent consumption.
5. SFE is an ideal technique to perform extraction of thermo-liable material as extraction can be performed at low temperatures.
6. SFE is an ecofriendly extraction technique which uses no or significantly less environmentally hostile organic solvents.
7. Direct coupling with chromatographic techniques is possible with SFE and can be a useful means to extract and directly quantify highly volatile compounds.
8. The supercritical fluid CO_2 which is used in SFE can be recycled or reused thus minimizing waste production.

9. SFE can be applied in different scales ranging from analytical scale (less than a gram to a few grams of samples), to preparative scale (several 100 grams of samples), to pilot plant scale (kilograms of samples) and up to large industrial scale (tons of raw materials).

Properties and advantages of CO_2 as a supercritical fluid.

1. It is chemically pure, non polar, stable solvent.
2. It is colorless, odorless and tasteless.
3. It can be easily removed, allowing simple product isolation by evaporation to 100% dryness
4. It is highly selective.
5. It is a tunable solvent whose density can be varied to control product solubility.
6. It is nontoxic and nonflammable.
7. It is an environmental friendly solvent.
8. Low temperature extraction conditions in SFE result in minimal degradation of volatile compounds (Table 6.4).

6.1.1.3.2 Factors to be Considered in SFE method Development

6.1.1.3.2.1 Solubility of Target Compound Before developing a method for SFE of botanicals, the feasibility of using SFE has to be ascertained. To determine the feasibility of SFE as a potential extraction technique the solubility of the target compound in supercritical carbon dioxide or other supercritical fluid of choice has to be determined. If the compound is poorly soluble in supercritical fluid(s), SFE is probably not the preferred extraction method. Solubility experiments to determine the effect of temperature and pressure (which in turn controls the density) on the solubility of the target compound in the supercritical fluid have to be performed simultaneously.

TABLE 6.4 List of Supercritical Fluids

Fluid	Critical Temp.(°C)	Critical Pressure(atm)
Carbon dioxide	31.3	72.9
Nitrous oxide	36.5	72.5
Ammonia	132.5	112.5
n-pentane	196.6	33.3
n-butane	152.0	37.5
Xenon	16.6	58.4

6.1.1.3.2.2 Cosolvents The effect of cosolvents on the solubility of the analyte of interest would have to be determined next. Based on the information obtained, a suitable cosolvent can be chosen and used for the extraction. The above information is often available in the literature for most compounds. However, in the absence of literature data, especially for those new research based novel molecular entities of unknown identity, solubility experiments needs to be performed.

6.1.1.3.2.3 Matrix (Plant Material) While the consideration of parameters such as flow, temperature, density, time, and modifiers may be fairly straightforward, perhaps the least predictable factor is that of matrix effects. The matrix has the analyte either in the exocellular or endocellular structures. For extraction of endocellular material where the analyte is absorbed into the internal cellular channels, stronger extraction conditions may be necessary. The matrix or powdered plant material also carries its own modifiers in the form of water, and/ or fats, oil and pigments. If the desired analyte is of polar nature, the water content of the matrix will facilitate the extraction. If the desired analyte is nonpolar, the moisture content will then inhibit extraction. The opposite effects are evident with more fats and pigments in the matrix. If significant variations of the matrix are expected, an extraction scheme must be developed with these extremes in mind.

6.1.1.3.2.4 Particle Size Sample particle size is a critical factor for a satisfactory SFE process. Large particles may result in prolonged extraction because the process can become diffusion controlled. In fact, pulverizing a sample into fine powder can speed up the extraction and improve the efficiency, but it may also cause difficulty in maintaining a proper flow rate.

6.1.1.3.2.5 Density Density is an important parameter since the solvating power of the supercritical fluid is proportional to its density. Higher the density, the more the analyte can be extracted from the plant matrix. Temperature control is necessary because the density of a supercritical fluid is directly related to the temperature for any given pressure.

6.1.1.3.2.6 Flow Flow is an important consideration because of the partitioning coefficients of the analyte between carbon dioxide and the plant matrix. Higher flows (or long extraction times) may be necessary to sweep all the analyte out of the extraction chamber. Low flows may be necessary if the kinetics of the system is slow.

6.1.1.3.2.7 Fluid Materials and Modifiers In order to extract polar compounds, polar supercritical fluid materials should be considered. Two such polar materials that have been successfully used for SFE of polar compounds from botanicals are Freon-22 (chlorodifluoromethane) and nitrous oxide. The former was used for extraction of free carboxylic acids and steroid compounds, while the latter for extraction of taxol. However, their application is limited due to their unfavorable properties with respect to safety and environmental considerations. Nitrous oxide can cause explosion and Freon-22 is no longer available commercially because of its ozone depletion effect in the upper atmosphere. Water as supercritical fluid material has also been reported. Superheated water (water under pressure and above 100 °C but below its critical temperature of 374 °C) has been used for extraction of herbal samples. Even though, superheated water has certain advantages such as higher extraction ability for polar compounds; it is not suitable for thermally labile compounds. If oxygen is not carefully purged, water at high temperature can be corrosive and might cause damage to extraction vessels.

The application of modifiers probably is the simplest yet the most effective way to obtain a desired polarity of CO_2 based fluids. Modifiers can be added to the extraction chamber, or to the supercritical fluid, to increase the polarity range for extraction. Pure carbon dioxide is useful for nonpolar to slightly polar compounds; therefore a modifier must be used to extract polar compounds. Commonly used modifiers are methanol, ethanol, tetrahydrofuran, chloroform, methylene chloride, formic acid, and dichloromethane. Usually, addition of a small amount of a liquid modifier can enhance significantly the extraction efficiency and, consequently, reduce the extraction time. At least 17 modifiers have been studied in SFE of natural products. Among all the modifiers, methanol is the most commonly used because it is an effective polar modifier and is up to 20% miscible with CO_2. It is generally believed that high percentages of methanol can disrupt the bonding between the solutes and plant matrices. Ethanol, though not as polar as methanol, may be a better choice in SFE of natural products because of its lower toxicity. Several reports have successfully employed ethanol as a modifier in SFE of a variety of organic compounds from plants. Depending on the properties of the samples and the desired compounds, the best modifier usually can be determined based on preliminary experimental results. For instance, in SFE of phenolic acids, methanol was a much more effective modifier than acetonitrile, acetone, water or dichloromethane. In SFE of paclitaxel and baccatin III, among methanol, ethyl acetate, dichloromethane and diethyl ether, dichloromethane was the most effective one for paclitaxel but diethyl ether was the best for baccatin III. In another report, 4% of methanol or chloroform did not result in any improvement on the recovery of santonin (a sesquiterpene lactone), but 4% of acetonitrile could increase the recovery from 38% to 85%, while water saturated CO_2 could further increase the recovery to 92%. By manipulating the types and ratios of liquid modifiers, one may obtain different extraction results.

There are three common ways to introduce a liquid modifier into the SFE system, using a second pump; using premixed fluids from a cylinder; and direct spiking. Compared with the other two methods, direct spiking a liquid modifier into the SFE cell is the simplest and the most economical method. It also creates less mechanical and reproducibility problems. However, if the spiking method is used, one must be very careful to make sure that the binary fluid is indeed in the supercritical state. Another common problem for the spiking method is that most of the modifier may be flushed out of the sample vessel in the very beginning of the dynamic extraction step, which will likely produce inconsistent results and, therefore will require repeated extractions.

One disadvantage of using a modifier is that it can cause poor selectivity, i.e. more impurity compounds such as chlorophylls and waxy material may be extracted with the desired compounds. If the concentration of a modifier is not properly selected, a lower recovery might be resulted probably due to the increased liquid modifier in the collecting solvent that can in turn produce a negative effect on the trapping efficiency.

6.1.1.3.2.8 Process Description A supercritical fluid extraction process consists of two steps: extraction of the components soluble in a supercritical solvent and separation of the extracted solutes from the solvent. The extraction can be applied to a solid, liquid or viscous matrix. Based on the objective of the extraction, two different situations may be considered.

1. Carrier material separation: In this case, the feed material constitutes the final product after undesirable compounds are removed. For example, dealcoholization of alcoholic beverages, removal of off-flavors, or decaffeination of coffee. However, this phenomenon is not so common in case of extraction of botanicals.
2. Extract material separation: In this case the compounds extracted from the solid feed material constitute the final product. For example, essential oil or antioxidant extraction.

The separation of soluble compounds from the supercritical fluid can be carried out by modifying the thermodynamic properties of the supercritical solvent or by an external agent. In the first case, the solvent power is modified by manipulating the operating pressure or temperature. In the second case, the separation can be carried out by absorption or adsorption, better called as solid trapping. The more common method decreases the operating pressure by an isoenthalpic expansion, which provides a reduction of fluid density and therefore a reduction of the solvent power. If separation takes place by manipulating the temperature, two situations may occur, depending on the solubility of the dissolved compounds. If solubility increases with temperature at constant pressure, a decrease in temperature will decrease the solubility and separate the compounds dissolved in the supercritical fluid. If solubility decreases with an increase in temperature at constant pressure, an increase in temperature will separate the compounds from the supercritical fluid solvent. Solid trapping is another common way for extract collection. Octadecylsilica (ODS) traps can be used at 15 °C to collect the essential oils and the bitter principles from hops, and after SFE the retained compounds can be rinsed out with acetonitrile for recovery. It was also found effective to retain volatile compounds on an ODS solid trap at lower temperature (15 °C) and then to elute these compounds with an organic solvent at an elevated temperature (45 °C). An advantage for solid trapping is that the selectivity can be further improved by selective trapping coupled with selective eluting. For example, polar compounds can be trapped on a silica gel column and then eluted with proper solvents. However, there are also some disadvantages for solid trapping methods. An improperly selected liquid modifier may result in significant loss of the desired compounds because they may be washed through the trap by the modifier solvent. Undesired reactions may also occur if the elution is done with improper solvents.

One of the main advantages of supercritical fluids is the ability to modify their selectivity by varying the pressure and temperature (by changing the fluid density). Therefore, supercritical fluids are often used to extract selectively or separate specific compounds from a mixture. In this regard fractional extraction process is very effective. In this process, the extraction is carried out in two stages. At the first stage, a relatively low fluid density is selected, which facilitates extraction of the compounds that are soluble at low pressure. Then, the residue is further extracted at high fluid density to recover heavier compounds. Another example of fractional extraction consists of removal of nonpolar fractions in the first stage by using a supercritical solvent and the removal of polar fraction from the residue in the second stage by adding a co-solvent.

Another procedure to selectively extract or separate specific compounds from a mixture is sequential depressurization. In this process, both fractions (light and heavy) are simultaneously extracted by using high density fluid. Then the supercritical solvent and the extract pass through multiple depressurization steps, allowing fractional separation. In the first depressurization stage, the heavier fraction is collected; the volatile or light fraction is collected in the last stage. Two depressurization steps are generally used. This method is commonly used in the extraction of spices, where the solubility of oleoresin and essential oil fractions in a supercritical solvent vary significantly with pressure and temperature. Generally, the extraction takes place at high pressures (40–60 MPa), so both fractions are soluble in the supercritical solvent. The separation or collection of the oleoresin fractions takes place in the first step by reducing the extraction pressure to intermediate pressure (15–20 MPa). Under such operating conditions, the aromatic fraction remains in the supercritical phase. After leaving the first separation, the pressure is further reduced and the essential oils are collected in the second step. This type of process has been successfully applied in multiple products. In some cases, both fractions are desirable (e.g., oleoresin and essential oil, color and pungent fraction), whereas in others, only one of the fractions has commercial value.

Figure 6.10 below shows a general flow diagram of a supercritical fluid extraction process from solids. The solvent is subcooled prior to the pump, assuring liquid phase to avoid cavitation issues. The pressurized solvent is heated above its critical temperature to the extraction temperature prior to the extraction vessel. The extraction vessel, which is filled with the feed material, is electrically heated or water heated to the extraction temperature.

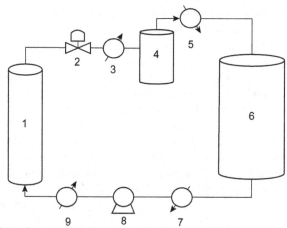

1: Extraction vessel, 2: Pressure reduction valve, 3: Vaporizer,
4: Separator, 5: Condenser, 6: Receiver, 7: Precooler, 8: Pump,
9: Preheater

FIGURE 6.10 General flow diagram of a supercritical fluid extraction process from solid material.

The supercritical solvent flows through the fixed bed and the soluble compounds are extracted from the carrier material. The supercritical fluid and the extract together leave the extraction vessel from the top, through a pressure reduction valve. The solvent power decreases with pressure reduction, so the compounds precipitate. To assure total complete precipitation, the supercritical solvent is heated above the saturation temperature to reach the gas phase. Under these conditions, the solvating power almost disappears. Then the material is collected in a separator while the solvent in gas phase leaves the separator vessel from the top and is re-circulated back to the extraction vessel. Once the raw material is fully extracted, the following steps are required in the extraction vessel:

• Depressurization
• Opening of the extraction vessel
• Unloading the spent material
• Loading with fresh material
• Closing the extraction vessel
• Pressurizing to operating conditions

Following are the types and techniques used in SFE:

6.1.1.3.2.9 Static and Dynamic SFE For dynamic extraction (SFE) the supercritical fluid is constantly flowing through the cell, and a flow restrictor is used to maintain pressure in the extraction vessel. In the case of static SFE, the cell is filled with the supercritical fluid, pressurized and allowed to equilibrate. After a period of time, a valve is opened to allow the analytes to be swept in the collection device.

6.1.1.3.2.10 Offline and Online SFE Techniques Offline SFE technique allows direct collection of the extracted analytes either in a solvent or passing the supercritical fluid through a column packed with chromatography adsorbent or collection in a cryogenic vessel for subsequent analysis. Online SFE technique, comprised of extraction and analysis, has been used by many researchers as an analytical tool for the study of medicinal plants. The extracted components are transferred directly from the SFE cell into the gas chromatography column via an on-column injector. SFE can be directly coupled with high performance liquid chromatography (HPLC) too with appropriate detector depending upon the type of analysis to be carried out. Offline SFE is simpler to perform and after extraction the extract can be analyzed by an appropriate analytical method. Online SFE requires an understanding of SFE and chromatographic conditions and the sample extract is not available for different analytical methods. However, sample handling between extraction and chromatographic analysis is eliminated by an online analysis. Loss of components can be avoided also. Whereas in offline SFE, the analytes are collected in a solvent and loss can occur during collection or handling during subsequent analyses (Table 6.5).

6.1.1.3.3 Advantages and Disadvantages of Supercritical Fluid Extraction (SFE)

6.1.1.3.3.1 Advantages

1. SFE offers vast array of possibilities for selective extractions and fractionations as the solubility of a chemical in a supercritical fluid can be manipulated by changing the pressure and/or temperature of the fluid.
2. Furthermore, supercritical fluids have a density of a liquid and can solubilize a solid like a liquid solvent.
3. The dissolved compounds can be recovered from the fluid by the reduction of the density of the supercritical fluid thus enabling solvent recovery.
4. Appropriate for heat sensible substances as supercritical CO_2 extraction uses a moderate extraction temperature as low as 30 °C.
5. More environmentally friendly extraction process than conventional solvent-solid extraction.

6.1.1.3.3.2 Disadvantages

1. Extraction of polar compounds cannot be done properly using Supercritical fluid.
2. Operating cost of SFE processes is too high and thus has restricted the applications to some very specialized fields such as essential oil extraction, coffee decaffeination and to university research only.

TABLE 6.5 Active Principles Isolated from Natural Sources by Supercritical Fluid Extraction

Source	Active Principle
Allium sativum	Organo sulfur compound
Catharanthus roseus	60 indole alkaloids detected
Cinchona succirubra	Quinine
Coffea arabica	Caffeine
Cotton seed	Cotton seed oil
Curcuma longa	Curcumin
Ginkgo biloba	Kaempferol, quercetin
Glycyrrizaglabra	Antimicrobial substance
Lavundula species	Terpenes
Mentha piperita	Mint oil
Morus alba	Triterpenoids
Papaver somniferum	Opium alkaloids
Piper nigrum	Pepper oil
Taxus brevifolia	Taxol
Silybummarianum	Polyphenolic compounds
Strawberry	Strawberry aroma
Lemon peel	Lemon peel oil

6.1.1.4 Accelerated Solvent Extraction/Pressurized Fluid Extraction/Enhanced Solvent Extraction

6.1.1.4.1 Theory

Accelerated solvent extraction (ASE) also known as pressurized solvent extraction, is a solid liquid extraction technique which has been developed as an alternative to current extraction methods such as Soxhlet, maceration, percolation or reflux, offering advantages with respect to solvent consumption, extraction yields, extraction time and reproducibility. ASE makes use of the same solvents as do other extraction techniques, but at an increased pressure (100–140 atm.) and at an elevated temperature (50–200 °C). The design of the extractor is such that it is capable enough to withstand high pressures, helps the extraction temperature to be raised above the boiling point of the solvent used. The high pressure helps in maintaining the solvent in a liquid state at a high temperature.

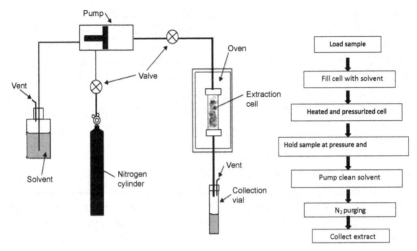

FIGURE 6.11 Schematic diagram of an accelerated solvent extractor along with step-wise sample loading and extraction steps.

Under such conditions, the solvent has properties favoring the extraction process, such as high diffusion coefficients, low viscosity and high solvent strength. This assists to attain a good dissolution processes and favors desorption of analytes from the cellular matrix. Additionally, pressure allows the extraction cell to be filled faster and helps to force liquid into the solid matrix. An increased temperature augments diffusivity of the solvent resulting in an increased extraction kinetics.

The sample is placed in an extraction cell (Figure 6.11), made of stainless steel. Following addition of the solvent, the cell is pressurized, heated to the desired temperature, and the sample is extracted statically for a specific period of time. Next, the extract is removed from the cell and the cell is flushed with fresh solvent. The cycle can be repeated. When the extraction is complete, compressed nitrogen moves all of the solvent from the cell to the vial for analysis. The extract is filtered prior to being collected in the receiver, thus eliminating the need for a separate filtration step. ASE has been used successfully for the extraction of analytes from medicinal plants, food, environmental samples, etc.

Types of ASE—Static and Dynamic mode of Extraction.

ASE can be performed in both static as well as dynamic mode, or as a combination of these two modes. In static ASE, the sample is extracted with a solvent at elevated temperature and pressure conditions without any outflow of solvent. When the extraction has reached equilibrium, the analytes are collected by rapidly flushing the extraction cell with solvent and an inert gas. In dynamic ASE, the extraction solvent is continuously flowing through the extraction cell. An advantage of dynamic ASE is that the solvent is continuously refreshed during the extraction. Evidently this technique requires a larger volume of solvent than static ASE and is, therefore, less suited for trace analysis. Static ASE, on the other hand, may lead to incomplete extraction because of the limited volume of solvent used. Dynamic ASE might be expected to yield faster extractions by continuously providing fresh extraction solvent to the sample, but this technique requires more solvent than static ASE, particularly when large samples are extracted. Therefore, a combination of static and dynamic extraction will often be the best choice in practice.

6.1.1.4.2 Factors Affecting ASE

6.1.1.4.2.1 Temperature of Extraction Elevated temperatures enhance the diffusivity of the solvent and thus resulting in increased extraction. Increasing temperature causes an increase in solubility of solutes, an increase in diffusion rates, a weakening of molecular interactions (van der Waals forces) between the solute and the matrix, and a decrease in the viscosity and surface tension of the solvent, which improves solvent penetration into the matrix leading to improved mass transfer.

6.1.1.4.2.2 Choice of Solvent In addition to extraction temperature, the extraction solvent is another important factor in ASE. Most ASE applications reported in the literature employed the organic solvents commonly used in conventional techniques such as methanol, in which many phytocompounds are readily soluble.

6.1.1.4.2.3 Pressure The utilization of elevated pressure allows solvents to be used above their boiling points to increase solvation power and extraction kinetics. These increase the extraction efficiency and rate, and reduce the consumption of organic solvents and operation time.

6.1.1.4.2.4 Nature of the Matrix The particle size of the matrix has an impact on the overall extraction yield. The surface area per unit mass of plant material increases as the particle size decreases, and this increases the solubility by promoting better solute-solvent interaction. In plant matrix, the passage of the analytes through the pores on the surface of the matrix also dictates the pace and efficiency of extraction.

6.1.1.4.2.5 Advantages and Disadvantages of ASE Use of nontoxic solvents such as carbon dioxide and water has socioeconomic and environmental benefits. ASE is considered to be a likely alternative technique to SFE for the extraction of polar compounds. Compared with traditional Soxhlet extraction, there is a remarkable decrease in the amount of solvent and the extraction time required for ASE. Meticulous attention should be paid to the accelerated solvent extraction performed with high extraction temperature, which may lead to degradation of thermolabile compounds.

6.1.1.4.2.6 Potential Applications of ASE Very few applications of ASE have been published in the field of natural product. There exists reports on the evaluation of ASE for the extraction of various metabolites covering a large range of structures and polarities (flavonolignans, curcuminoids, saponins, terpenes) present in different plant parts such as leaves, roots, fruits, herbs and rhizome. Yields were found to be equivalent or even higher with ASE, with a reduction in the solvent consumption, extraction time and with a good reproducibility most likely due to the minimal sample handling required during the extraction process.

6.1.1.5 Pulse Electric Field Assisted Extraction

Exposing a biological cell (either plant or animal) to a high intensity electric field (kV/cm) in the form of very short impulses induces the formation of temporary or permanent pores on the cell membrane. This phenomenon is termed as electroporation which actually increases the permeability of the cell under treatment, ultimately results in cell membrane disintegration. Before moving further a conceptual idea about electric field induced cell damage resulting in drug extraction has been put forward. Secondary metabolites located in the inner glands of the plant cells are extracted applying electric field through pore generation in the cell membrane.

The cell membrane of most living organisms consists of two phospholipids layer popularly known as phospholipid bilayer as shown in Figure 6.12. The lipid bilayer is oil like film consists almost entirely of cholesterol and phospholipid. The outer part of the bilayer is hydrophilic (water loving) and the inner part is hydrophobic (water repellant or fat soluble). Because the fatty portion are repelled by water but are mutually attracted to each other, they tend to line up occupying the center portion of the bilayer, while the hydrophilic portion face the water surrounding the cell both inside and outside. The membrane lipid bilayer is a major barrier

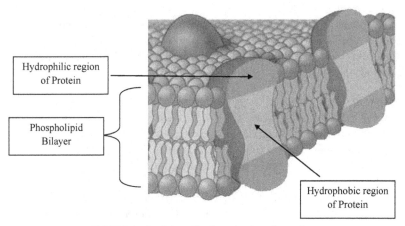

Hydrophilic region of Protein

Phospholipid Bilayer

Hydrophobic region of Protein

FIGURE 6.12   Schematic diagram of a cell membrane.

impermeable to water soluble entities such as ions, glucose and urea. Some proteins known as "Integral" proteins which floats in the lipid bilayer and protrudes through the membrane help in the formation of channels through which water soluble substances, especially ions can diffuse. By using these same pores water soluble drug entities placed inside the cell can come out once they are forced to do so.

6.1.1.5.1 Mechanism Behind Induction of Electroporation

It is a well known fact that outside of the cell bears a positive (+ve) charge while the inside of the cell has negative (−ve) charge due the presence of organic acids. Thus the distribution of ions in such a fashion across the cell membrane results in the formation of a transmembrane potential of 10 mV. The main cause of cell membrane destabilization is attributed to the increase of the so-called transmembrane potential and the phenomenon of electroporation can be described as a dramatic increase in membrane permeability caused by externally applied short and intense electric field. Various theoretical models were developed to describe electroporation among which the Transient Aqueous Pore model is most widely accepted.

6.1.1.5.1.1 Aqueous Pore Model According to this model when a cell is exposed to an external electric field, the induction of an induced transmembrane potential provides the free energy necessary for structural rearrangements of membrane phospholipids and thus enables hydrophilic pore formation. On the other hand hydrophobic pores are formed by spontaneous thermal fluctuations of membrane lipids. Using these pores both water soluble and fat soluble drug molecules find exit routes to come outside.

6.1.1.5.1.2 Electro-Compression Model According to Zimmermann, the introduction of an external electric field leads to an enhanced attraction of opposite charges existing outside (+ve) and inside (−ve) of the cell and results in electro-compression of the cell membrane ultimately resulting in pore formation as shown in Figure 6.13.

The theoretical description of the transmembrane potential induced on a spherical type of cell exposed to electric field is known as Schwan's equation. The induced transmembrane potential for a spherical cell can be calculated as shown in Equation (1.2):

$$U_{TI} = -1.5rEcos\varphi \qquad\qquad 1.2$$

where r is the radius of the cell, E is the strength of applied electric field, and φ is the angle between the direction of the electric field and the selected point on the cell surface. The maximum electroporation occurs at the poles of the cell exposed to the electric field facing the electrodes when electric field passes from anode to cathode.

Pulse electric field (PEF) technology is based on electric power delivered to the product placed between two electrodes. The equipment consists of a high voltage pulse generator and a treatment chamber with necessary monitoring and control devices. The raw materials are placed in the treatment chamber as shown in Figure 6.14, where two electrodes are connected together with a nonconductive material to avoid electrical flow from one to the other. Generated high voltage electrical pulses are applied to the electrodes, which then carry the high intensity electrical pulse to the sample placed between the two electrodes. The sample experiences a force per unit charge (the electric field), which is accountable for the irreversible cell membrane damage and breakdown. Thus depending upon the applied electric field strength, the cells can be either non permeabilized, when the so-called critical electric field strength (E_{crit}, it is the minimal field strength required to induce pore formation) is not exceeded, reversibly permeabilized if the E_{crit} value is slightly exceeded and irreversibly permeabilized as soon as the E_{crit} is extensively exceeded as shown in Figure 6.15. The secondary metabolites which are located deep inside the cell matrix now have the access of the dissolving solvents which in this case is termed as electroporating medium.

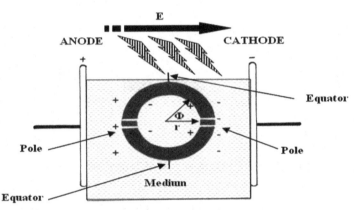

FIGURE 6.13 The induced transmembrane potential is maximal at the poles of the cell. Electroporated area is presented as noncontiguous line at the poles.

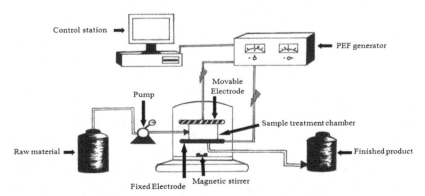

FIGURE 6.14 Continuous pulse electric field (PEF) processing.

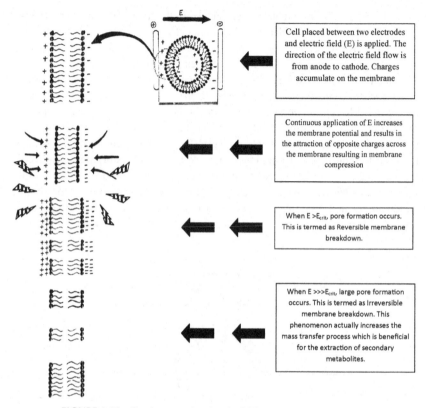

FIGURE 6.15 Continuous pulse electric field (PEF) processing system.

6.1.1.5.2 Applications of Pulsed Electric Field Assisted Extraction

Based on the above mentioned phenomenon, electroporation (permeabilization of the cell membrane caused by external electric field) was applied in the field of secondary metabolite extraction.

6.1.1.5.2.1 Extraction of Secondary Metabolites in Plant Cell Cultures
Following is the list of several bioactives which were extracted using PEF technique (Table 6.6).

6.1.1.5.2.2 Juice Extraction
Pulsed electric field has been largely used in the extraction of juice and anthocyanins from grapes and grape byproducts. Expression and characteristics of juice extracted from white grapes was also studied whereby an increase in juice yield from 49% to 76% was noticed. An improvement in juice yield from sugar beet slices treated at 215, 300, and 427V/cm at 500 pulses showed increased yield of 43%, 68%, and 79% for the above electric fields. The influence of PEF treatment on apple mash was also investigated using field

TABLE 6.6 Secondary Metabolites Extracted Using Pulsed Electric Field Assisted Extraction

Secondary metabolites	Source
Amaranthine	Cultured plant tissue of *Chenopodium rubrum*
Anthraquinones	Cultured plant tissue of *Morinda citrifolia*
Betalain	Beetroot tubers of *Beta vulgaris*
Carotenoids	Carrots
Vitamin C, β-carotene	Red bell peppers
Taxuyunnanine	Taxus
Anthocyanins	Grapes
Antioxidants	Apple mash

strength of 0.5–5 kV/cm, applying 10–40 pulses. After a treatment at 3 kV/cm and 20 pulses a juice yield of 83.0% was obtained after pressing in comparison to 80.1% after enzymatic maceration and 75.4% for the untreated control.

Momentous steps forward have been made for pulsed electric field processing and it has reached a point where it is very close to marketable realization. Conventional and novel extraction techniques actually depend on expensive extraction parameters ranging from costly solvent to extraction energy involved to attain complete extraction. Moreover environmental related issues are also an area of great concern when extractions of botanicals are concerned. As a secondary, but very interesting attributes are its nonthermal characteristic and its ability to preserve the integrity of the extracted products. This could become, with time, the main key for the success of the application of PEF techniques to the extraction of natural biologically active components from plants. Harmonization of equipment and research protocols is taking place and this will greatly help the situation. As research progress is made and knowledge increases regarding the most effective design parameters for extraction yield whilst minimizing product deterioration, then the full potential of PEF may be realized in the near future.

6.1.1.6 Steam Distillation

It is a valuable technique, which allows isolation of volatile components, such as essential oils, amines, and organic acids, as well as other, relatively volatile compounds insoluble in water. Among others, steam distillation has been used to isolate the essential oil fraction from either plant material or previously prepared extract in a low boiling solvent such as petroleum ether. The plant

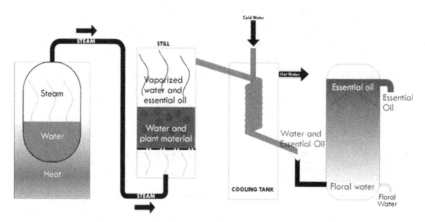

FIGURE 6.16 Schematic description of steam distillation process.

material is generally loaded into a cylindrical still with perforated bottom from which steam rises up (shown in Figure 6.16). The steam saturated with the volatile component of the plant material is then passed through a condenser. The oil and water slowly separates out on standing or by centrifuging.

However, the technique is not free from shortcomings as it involves substantial energy consumption. Usage of elevated temperature may cause thermal decomposition of the analytes of interest. This can also affect the essential oil components resulting in flavor changes.

6.1.1.7 Solid Phase Micro Extraction

This is a novel effective solvent less sample preparation technique best suited for gas chromatography (GC). Solid phase microextraction (SPME) uses a short length of narrow diameter a fused silica optical fiber externally coated with a thin film polymeric (e.g. carbowax, polydimethylsiloxane (PDS)) stationary phase or a mixture of polymers blended with a porous carbon based solid material (e.g. PDMS-Carboxen).

The coated fiber is immersed directly into the sample, where analytes preferentially partition by adsorption or by absorption from the solution to the stationary phase (Figure. 6.17). After equilibrium is reached, which can range from few minutes to several hours depending upon the chemical nature of the analyte of interest, after a definite time, the fiber is withdrawn and introduced into the injection chamber of a gas chromatograph, or it is introduced into a modified HPLC rheodyne valve. The fiber is exposed and the analytes desorbed either thermally into the hot GC injector or, in the case of HPLC, eluted by the mobile phase, and subsequently conventionally chromatographed. The advantages of SPME includes.

- Detection limits down to parts per trillion can be obtained.
- SPME is solvent free and the fiber can be used repeatedly.

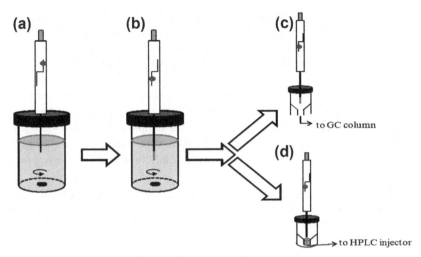

FIGURE 6.17   Configuration of solid phase micro.

Hence, SPME can be successfully used for the characterization and quantification of the biogenic volatile organic compounds (e.g. isoprene and terpenoid compounds) present in plants. SPME-GC has also been used for the determination of tobacco alkaloids. Extraction and equilibrium process in case of SPME can be varied and enhanced in different ways. While extracting semivolatile compounds from an aqueous matrix the fiber is usually immersed directly into the solvent. If the sample is agitated with magnetic stirrer or ultrasonically, the time to reach the equilibrium can be lowered significantly. Time to attain equilibrium also depends upon the analyte nature and fiber chemistry. Extraction efficiency can be improved by modifying matrix properties, target analytes and the SPME device itself. To achieve precision and reproducibility these conditions and others such as incubation temperature, sample agitation, sample pH and ionic strength, sample volume, extraction and desorption times must be kept robust.

6.1.1.8 Head Space Analysis (Static Headspace and Dynamic Headspace i.e., Purge and Trap Sampling)

For classical GC generally samples must be dissolved in organic solvents injections and, for modern capillary GC, be dilute, some sample preparation is required in nearly all GC analytical methods. Because GC requires that samples be volatile, it is a natural tool for the analysis of vapor-phase mixtures. Headspace analysis is generally defined as a vapor-phase extraction, involving the partitioning of analytes between a nonvolatile liquid or solid phase and the vapor phase above the liquid or solid. Volatile organic compounds can be concentrated by either static headspace or dynamic headspace (i.e., purge and trap) sampling.

In static headspace concentration, a sample is placed in a closed sample chamber. Molecules of the volatile compounds in the sample migrate to the headspace above the sample and equilibrium is established between the concentration of the compounds in the vapor phase and in the liquid phase (Figure 6.18).

Once equilibrium is reached, an aliquot of the headspace above the sample is injected onto the GC column. A major problem with static headspace techniques is that the sample matrix significantly affects equilibrium. Analyses for compounds that show high solubility in the sample matrix often yield low sensitivity as a result of matrix effects. Further, static headspace analysis only samples an aliquot of the volatiles (i.e., 1 mL, 2 mL, or whatever the size of the sample loop), which also affects sensitivity.

The sample phase contains the compound(s) of interest, usually in the form of a liquid or solid in combination with a dilution solvent or a matrix modifier. Once the sample phase is introduced into the vial and the vial is sealed, molecules of the volatile component(s) diffuse into the gas phase until the headspace reaches a state of equilibrium, as depicted by the arrows. An aliquot is then taken from the headspace.

Purge and trap concentration is a dynamic headspace technique that reduces matrix effects and increases sensitivity, relative to static headspace techniques. Samples containing volatile organic compounds are introduced into a purge vessel and a flow of inert gas is passed through the sample at a constant flow rate for a fixed time. Volatile compounds are purged from the sample into the headspace above the sample and are transferred to and concentrated on an adsorbent trap. After the purging process is complete, the trap is rapidly heated and back flushed with carrier gas to desorb and transfer the analytes to the GC column.

6.1.1.9 Sample Disruption Method

In the recent time, matrix solid-phase dispersion (MSPD) has become an extraction method for naturally occurring compounds. The most significant features

FIGURE 6.18 Volatile analyte in equilibrium between the gas and sample phase **G** = gas phase, **S** = sample phase.

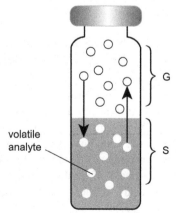

volatile analyte

of this technique are its selectivity, flexibility and the possibility of performing extraction and clean-up in one step. This technique is based on blending of a viscous, solid, or semisolid sample with an abrasive solid support material. The detailed steps of the extraction technique are given below.

1. The solid/liquid/semisolid extract is blended together with a solid support, using a glass mortar and pestle, to obtain complete disruption and dispersion of the sample on the solid support.
2. The sample is packed into an empty column or on top of a solid phase extraction sorbent. However, the main difference between MSPD and SPE is that the sample being dispersed throughout the column and not retained in just first few millimeters of the column.
3. Elution can be performed in two ways. In the first method the target analytes are retained on the column and interfering compounds are eluted in the washing step, while the target analytes are subsequently eluted by a different solvent. In the second method the interfering matrix components are selectively retained on the column and the target analytes are directly eluted.
4. Sample can be directly subjected to chromatographic analysis.

6.1.1.10 Extraction with Superheated Water

Water is a unique solvent because of being highly hydrogen bonded structure, and at room temperature it has a disproportionately high boiling point. However, when heated the properties of water changes markedly. With the increase in temperature, there is a marked and systematic decrease in the permittivity, an increase in the rate of diffusion and a decrease in the viscosity and surface tension. If the pressure is raised so that the water remains in a condensed state, these changes continue above the atmospheric boiling point and up to and beyond the critical point at 374 °C and 218 bar. Over much of this temperature range the density is almost constant so that pressure effects on the properties of water are minimal. In this case superheated water is used as a general term to denote the region of the condensed phase between 100 °C and the critical point. The pressure required to maintain a condensed state of water is 15 bar at 200 °C and 85 bar at 300 °C. As the temperature of liquid water is raised under pressure, between 100 and 374 °C, the polarity decreases significantly and it can be used as an extraction solvent for a wide range of analytes. Studies on plant materials have concentrated on two areas, the extraction of naturally occurring plant products, principally essential oils and secondly on the determination of pesticide residues. In SWE (extraction with superheated water) the need for the plant material to be dried before extraction can be eliminated. Superheated water extractions have been shown to be feasible with particular interest in avoiding the need for organic solvents. The method is thus environment friendly, cheap and nontoxic. One disadvantage of SWE is that the extract is relatively dilute aqueous solution and this raised concerns about the solubility of analytes and potential for precipitation and sample loss by re-adsorption onto the original matrix. This is principally a problem when there are marked differences in solubility on cooling. The dilute extract although free of the matrix often needs a concentration step before any subsequent assay step. However, because the extract solution is a clean matrix, sample handling and concentration is much easier than from the original sample material. The aqueous extraction solution can be passed through a solid-phase extraction cartridge and then the analytes can be extracted by solvent elution in a small volume before subjecting to chromatography. Because of the much lower solubility of very nonpolar analytes in cold water there is a danger that they might be deposited back onto the sample matrix. A successful method in these cases has been the addition of a trapping agent (a styrene divinyl extraction disc) to the extraction vessel so that the extracted materials are trapped out of the cooling aqueous phase within the extraction vessel. As the aqueous solution cools down the analytes are partitioned into the disc.

6.1.2 Preparative Extraction Methods

6.1.2.1 Soxhlet Extraction

Soxhlet, which has been used for a long time, is a standard technique and is the main standard for evaluating the performance of other solid–liquid extraction methods. Although Soxhlet extraction in general is a well established technique, but suffers from few drawbacks.

6.1.2.1.1 Mechanism of Action

In a Soxhlet extractor the sample is placed in a thimble-holder which during the whole operation is been gradually washed by fresh solvent or solvent combination through evaporation from a distillation flask. When the liquid reaches the overflow level, a siphon aspirates the extracting solvent of the thimble-holder and unloads it back into the distillation flask, carrying the extracted analytes into the bulk liquid as shown in Figure 6.19. This operation is repeated until complete extraction is achieved. As the solvent in a Soxhlet proceeds stepwise, the assembly can be termed as a batch design; however, as the solvent is recirculated through the sample matrix, the system also bears a continuous behavior.

6.1.2.2 Factors Determining Soxhlet Extraction

6.1.2.2.1 Solvent Choice

A suitable extracting solvent should be selected for the extraction of targeted bioactive compound using the Soxhlet extraction method. Generally the most widely used solvent to extract oily or fatty materials from plant matrices are the less polar solvents. Moreover it has been seen that methanol, ethanol, and water exhibit similar solubility properties since they contain a hydroxyl group which is hydrophilic in nature.

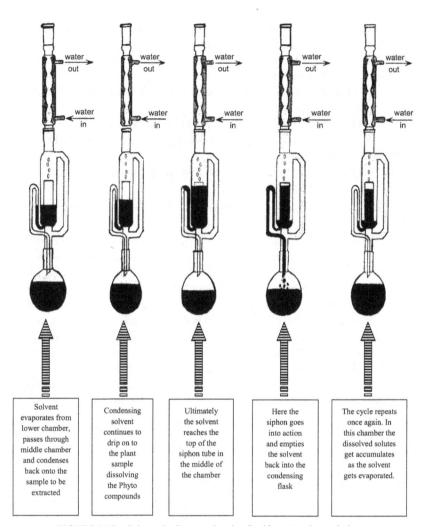

Solvent evaporates from lower chamber, passes through middle chamber and condenses back onto the sample to be extracted	Condensing solvent continues to drip on to the plant sample dissolving the Phyto compounds	Ultimately the solvent reaches the top of the siphon tube in the middle of the chamber	Here the siphon goes into action and empties the solvent back into the condensing flask	The cycle repeats once again. In this chamber the dissolved solutes get accumulates as the solvent gets evaporated.

FIGURE 6.19 Schematic diagram showing Soxhlet extraction technique.

Petroleum ether, acetone, chloroform, n-hexane, ethyl acetate might be the best solvents to eradicate the unwanted oily matter. From safety point of view, the use of excess amount of unsafe organic solvents may not be a good choice for the processing of highly valued phytochemicals, especially for the use in the pharmaceutical industry. So such considerations are very vital while selecting solvent or solvent systems for Soxhlet extraction.

The following factors should be considered while selecting a solvent for Soxhlet:

- Selectivity—Only the active, desired constituents should be extracted from the plant material, which means that a high selectivity is required.

- Boiling temperature—The boiling point of the solvent should be as low as possible in order to facilitate easy solvent recovery.
- Reactivity—The solvent should not react chemically with the extract, nor should it readily decompose.
- Safety—The solvent should be nonflammable and noncorrosive, and should not present an environmental risk.
- Cost—The solvent should be readily available at low cost.
- Vapor pressure—Must possess a low vapor pressure to prevent solvent loss during extraction.
- Recovery—The solvent has to be separated easily from the extract to produce a solvent-free extract.

6.1.2.2.2 Matrix Characteristics

Knowledge of the matrix characteristics of the carrier solid is important to determine whether it needs prior treatment to facilitate the diffusion of solvent to occur. Solute may exist in the inert solids in a variety of ways. Either by adhering on to the solid surface or being surrounded by a matrix of inert material or inside the cells. Solute adhering to the solid surface can be readily removed by the solvent. But when the solute exists in pores surrounded by a matrix of inert material, the solvent has to diffuse into the interior of the drug matrix to solubilize the analyte. So to facilitate the entry of solvent herbal industries generally use ball mills and fluid mills for size reduction to obtain particles with greater surface areas which facilitates better diffusion of solvent resulting in improved extraction yield. In a study it has been found that for the extraction of total fat from oleaginous seeds, a 2-h extraction obtained 99% extraction efficiency when the particle size was 0.4 mm, while a 12-h extraction was necessary to obtain similar efficiency if the particle size was near to 2.0 mm.

6.1.2.2.3 Volume of the Extracting Solvent

Volume of the extracting solvent is also a critical factor. The overall knowledge is that the solvent volume must be sufficient to ensure that the plant matrix is always entirely immersed in the solvent throughout the entire extraction process.

6.1.2.2.4 Advantages and Disadvantages of Soxhlet extraction

The advantages of conventional Soxhlet extraction include (1) the displacement of transfer equilibrium by repeatedly bringing fresh solvent in contact with the solid matrix (2) maintaining a relatively high extraction temperature and (3) no filtration requirement after leaching. Moreover, the Soxhlet method is very simple and cheap.

The main disadvantages of conventional Soxhlet extraction includes (1) prolonged extraction time; (2) large volume of solvent being used; (3) agitation cannot be provided in the Soxhlet device to accelerate the process; (4) the large amount of solvent is used; and (5) the possibility of thermal decomposition of the target compounds cannot be ignored as the extraction usually occurs at the boiling point of the solvent for a long time. The long time requirement and the requirement of large amounts of solvent lead to wide criticism of the conventional Soxhlet extraction method.

6.1.2.2.5 Trivial Soxhlet Adjustments

The majority of simple modifications developed within the original Soxhlet device have been done to maximize the extraction yield when thermolabile and liquid samples are being used. These changes have been incorporated as minor alterations of basic units such as the thimble-holder, siphon, condenser, etc. which slightly have improved the features and results of the methods thus developed. A sealed Soxhlet extractor has been designed for extraction of lipids under constant moisture and vacuum conditions.

For thermolabile substances the extraction thimble-holder has been designed away from solvent flask as shown in Figure 6.20 above in such as way that the heat from the heating source never reaches the sample.

As shown in Figure 6.21 below the siphon has also been modified by putting a sintered-glass disc at the bottom of the extraction chamber and an outlet with a teflon stopcock below the disc. The stopcock has twin purpose of controlling the flow of the solvent and maintaining a constant solvent level above the solid sample to be extracted.

6.1.2.2.6 Major Soxhlet Improvements

Most of the modifications of the conventional Soxhlet extractor developed in the last few years have been aimed at making its performance more similar to

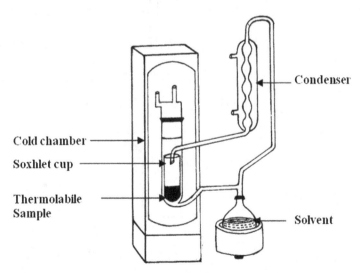

FIGURE 6.20 Modified Soxhlet extractor for the continuous extraction of thermolabile substances.

that of the recent techniques for solid sample treatment namely shortening the time for the leaching step through automation. Thus, the most essential alterations of conventional Soxhlet have led to the design of automated and microwave-assisted Soxhlet extractors along with other major changes.

6.1.2.2.7 Ultrasound-Assisted Soxhlet Extraction

The perfect advancement would be the one retaining the advantages of Soxhlet extraction (namely, sample-fresh solvent contact during the whole extraction step, no filtration step, simple manipulation) while overlooking its shortcomings by accelerating the process and minimizing environmental hazards. The modification is based on the same principle as in conventional Soxhlet extractor but has been modified to facilitate the location of the Soxhlet chamber in a thermostat bath through which ultrasounds are applied by means of an ultrasonic horn as shown in Figure 6.22. The extraction process consisted of a number

FIGURE 6.21 A Soxhlet extraction unit with a Teflon stopcock for sampling.

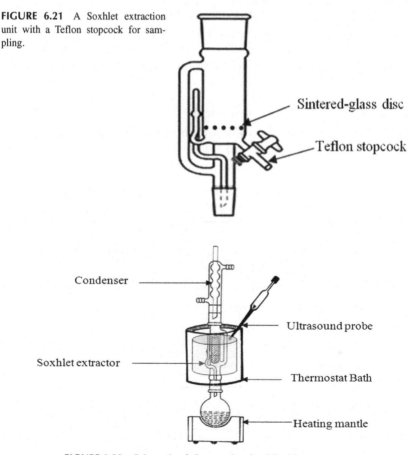

Sintered-glass disc

Teflon stopcock

Condenser

Ultrasound probe

Soxhlet extractor

Thermostat Bath

Heating mantle

FIGURE 6.22 Schematic of ultrasound-assisted Soxhlet extractor.

of cycles that depends on the extraction kinetics of the target sample. Each cycle involved three steps: (1) filling of the Soxhlet chamber by the solvent evaporating from the distillation flask followed by condensation and dropping on the sample, (2) ultrasound irradiation of the sample, (3) unloading of the Soxhlet chamber content after the extraction reached the siphon height. After the last cycle, the solvent from the extracted materials are removed by using rotary evaporator.

6.1.2.2.8 Automated Soxhlet Extraction

This technique is an automated version of the classic Soxhlet approach for extracting solid samples and was proposed by Edward Randall in 1974. The system works in the following ways.

First, the thimble is lowered into the boiling solvent until the appropriate extraction takes place. Later on, to flush residual extract from the sample, a rinse step follows which is step two as shown in Figure 6.23.

In this second stage, the thimble is raised above the boiling solvent for a small time until residual extract is removed from the solid mass by the condensed solvent, like in an original Soxhlet experiment. Finally, through a drying step the solvent is removed from the solvent flask and the analyte is concentrated later for further processing. In some other type of systems, there is a fourth step in which the sample holder is lifted from the heat source and is allowed to evaporate further without giving any chance of sample overheating or potential oxidation. This approach can decrease the extraction time by as much as a factor of 10 compared with traditional Soxhlet extraction.

6.1.2.2.9 Microwave-Assisted Soxhlet Extraction

In the midst of the various attempts to improve Soxhlet performance, the most successful has been the use of microwave, which has provided the wider variety of approaches. In fact, microwave-assisted Soxhlet extraction remains the most interesting improvement of conventional Soxhlet extraction. Microwave-assisted Soxhlet extraction differs mainly in some aspects from other microwave-assisted extraction techniques, namely (1) the extraction vessel is open, so it always works under normal pressure; (2) microwave irradiation is focused on the sample compartment; (3) continuous sample–solvent contact (4) no subsequent filtration is required. Therefore, these approaches retain the advantages of conventional Soxhlet extraction while overcoming its limitations.

The two alternatives of microwave-assisted Soxhlet are the Soxwave-100 extractor, and the FMASE.

6.1.2.2.9.1 Soxwave-100 The principle of Soxwave-100 is similar to Kumagawa extraction (which is very similar to the Soxhlet extractor. The Kumagawa extractor has a particular design where the thimble chamber is directly suspended inside the solvent flask which is having a vertical large opening above

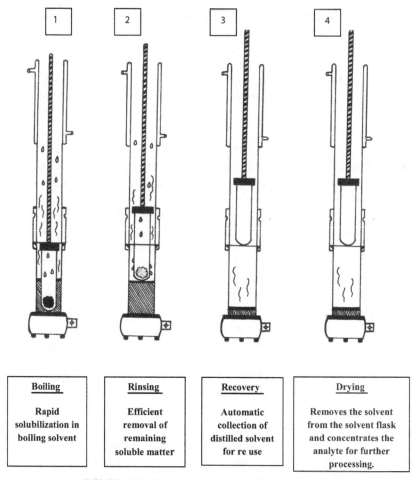

Boiling	**Rinsing**	**Recovery**	**Drying**
Rapid solubilization in boiling solvent	Efficient removal of remaining soluble matter	Automatic collection of distilled solvent for re use	Removes the solvent from the solvent flask and concentrates the analyte for further processing.

FIGURE 6.23   Working of an automated Soxhlet extractor.

the boiling solvent). The thimble is surrounded by vapor of hot solvent and is maintained at a higher temperature compared to Soxhlet extractor, thus assisting in better extraction for compounds with higher melting points. The removable chamber is clamped with a small siphon side arm and, in the same way as for Soxhlet; a vertical condenser makes sure that the solvent falls back down into the chamber which is automatically emptied at every cycle thus following a three step extraction. Moreover the Soxwave-100 extractor uses a single heating source (i.e., focused microwaves) which acts both on the sample and the solvent.

6.1.2.2.10 Focused Microwave-Assisted Soxhlet Extractor

The FMASE device as shown in Figure 6.24 works as a conventional Soxhlet; that is, it also follows a series of cycles in which the extractant is completely

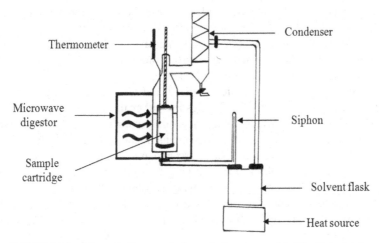

FIGURE 6.24 Schematic diagram of a focused microwave-assisted Soxhlet extractor.

renewed but with sample irradiation by microwaves at preset time within each cycle. The FMASE generally uses two energy sources: microwaves for sample irradiation and electrical heating of the extractants. The following modifications allow some benefits which are as follows. (1) sample heating is independent of the solvents polarity; (2) the energy for solvent heating required to remove the target analytes from the sample can be optimized independently; (3) microwave irradiations are simultaneous, which facilitates mass transfer and shortens extraction times in the process.

Focused microwave-assisted Soxhlet extraction provides the following advantages over the conventional method of extraction: (1) substantial shortening of the extraction time; (2) saving of extracting solvent; (3) no moisture adjustment.

6.1.2.3 Maceration/Simple maceration—A Steady-State Extraction

Before the nineteenth century, there was no real progress in methods and techniques of extraction from plant materials for industrial use.

The techniques available were simple expression, aqueous extraction and evaporation followed by the use of extraction processes using alcohol as a solvent. Soon such techniques became highly successful in the phytochemical extraction resulting in the isolation of single pure molecules for industrial and medicinal purpose. Above mentioned techniques were primarily used for manufacturing various classes of medicinal plant preparations, such as decoctions, infusions, fluid-extracts, tinctures, semisolid extracts and powdered extracts, popularly known as *galenicals*. The sole purpose of those extraction techniques was to obtain the therapeutically desirable portion by eliminating the inert material by treating with a particular solvent.

Maceration is preferably used with volatile or thermal instable products. It is "a cold" extraction of pulverized feed material in any solvent. In due course of time maceration became a popular and inexpensive way to get essential oils and bioactive compounds. For small scale extraction, maceration generally consists of several steps as shown in Figure 6.25 below.

Maceration process for both organized and unorganized drugs varies as shown in Figure 6.26. Organized drugs are either the parts of the plants like roots, seeds and barks which contain secondary metabolites (alkaloids, glycoside, tannins, steroids etc.) as their main constituents. On the other hand unorganized drugs are derived from plant parts and are noncellular by nature. Their main constituents are volatile oils, oleo-gum resins, resins etc. In maceration for organized and unorganized drugs following steps are followed:

In the above mentioned single maceration process, some considerable amount of the liquid may be left behind after pressing the marc for the first time. As this liquid also contains bioactive constituents at a concentration equal to the strained liquid, repeated maceration will be more efficient when the bioactives are more valuable. Sometimes double maceration is used for concentrated infusions containing volatile oils. In cases where the marc cannot be pressed, triple maceration is employed. Kinetic or dynamic maceration, turbo extraction, infusion, decoction and digestion are some of the derivatives of maceration method.

In Kinetic or Dynamic maceration both the herbal drug and the solvent are maintained under constant stirring. In today's time a new dynamic maceration

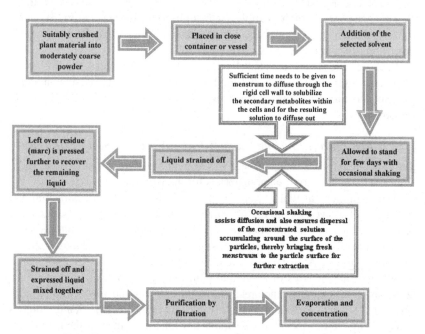

FIGURE 6.25 General laboratory procedure for a maceration process.

technique using high CO_2 pressure during the maceration is commonly under use especially in wine industries. Pressure differences used in this technique could significantly affect the cell anatomy, resulting in a better extraction of polyphenols and aroma precursors. Such maceration can result in wines with higher anthocyanin content and better color intensity, which is very important in the red wine technology.

Another subdivision of maceration is turbo extraction where the extraction is generally conducted using cold solvent under high shear forces (50,000–2,00,000 rpm). Such technique simultaneously results in particle size reduction, cell structure disintegration and temperature rise.

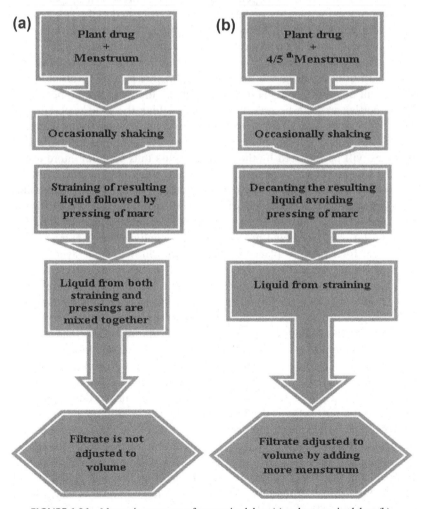

FIGURE 6.26 Maceration processes for organized drug (a) and unorganized drug (b).

Infusion and Decoction are also the derivatives of maceration whereby water is used as an extracting solvent at varying temperature condition. In infusion, boiling water is placed over the plant material to be extracted and is allowed to stand for few minutes followed by filtration. Whereby in decoction the plant materials are placed in cold water and then heated to induce boiling for 10–15 min if possible under reflux to prevent loss on evaporation. This technique finds its usefulness in case of those herbal drugs which are thermostable.

6.1.2.4 Percolation–An Exhaustive Extraction Process

When the mean diameter of the feed material increases percolation is a widely used technique. It is generally used for unorganized drugs. In percolation process a suitably powdered form of drug material is packed in a percolator and the solvent or menstrum is allowed to percolate through it. If a too dry material is packed into the percolator, it will immediately swell and will increase in volume due to solvent absorption and thus affect the process of drug leaching. On the other hand if fine powder material is used it will run down the percolator and will choke the pores of the percolator. These bottlenecks can be prevented by preleaching the plant materials with the solvent or the menstrum prior to extraction and the process is termed as *imbibition*. During imbibition the air or air bubbles entrapped within the drug particles will be replaced by the solvent vapors thus enabling more evenly packed drug bed and easy menstrum flow. After attaining a suitable hydration status the drug materials are packed evenly into the percolator. The imbibed drug is packed over a loose plug of tow or other suitable material previously moistened with the solvent. Even packing can be achieved by introducing the material layer by layer and pressing it occasionally. After packing is over, a piece of filter paper is placed on the surface followed by a layer of clean sand such that the top layer of the drug is not disturbed when solvent is added for extraction. Sufficient solvent is now poured over the drug sample slowly and evenly to soak it, keeping the tap at the bottom open to allow the gases between particles to pass out. Solvent should never be poured with the tap closed since the occluded air will escape from the top, disturbing the bed. When the solvent begins to drip through the tap, the tap is closed; sufficient solvent is added to maintain a small layer above the drug sample and allowed to stand for 24 h. The layer of solvent above the surface of the sample bed prevents drying of the top layer, which may result in the development of cracks on the top surface of the bed. The 24 h maceration period allows the solvent to diffuse through the drug sample, solubilize the constituents and leach out the soluble material. After the maceration, the outlet is opened and the solvent is percolated at a controlled rate with uninterrupted addition of fresh solvent. At first only 75% of the volume of the finished product is collected after which the marc is pressed and expressed liquid is added to the percolate giving 80–90% of the final volume.

While preparing concentrated products after evaporation few problems may arise.

- When the plant material is thermolabile (evaporation may result in partial loss of the active constituents).
- When the menstrum is a combination of water and alcohol (vaporization of alcohol will take place leaving behind an almost aqueous concentrate which may not be able to retain the extracted matter in solution and hence may get precipitated).

Both the above mentioned problems can be overcome by reserving the first portion of the percolate containing the bulk of the active constituents. The rest of the percolate is collected separately. This second dilute percolate is evaporated under reduced pressure followed by evaporation until a soft extract is attained so that almost all the water is removed. It may then be dissolved with the reserved percolate which is strongly alcoholic. This process is called reserved percolation which actually prevents precipitation.

6.1.2.5 Enfleurage

Regardless of the introduction of the modern extraction process for volatile components, the old fashioned technique of enfleurage still plays an important role. The principle of enfleurage is quite simple. An inherent property of fat which is used in this technique is that it possesses a high power of absorption and, when brought in close contact with fragrant flowers, it readily absorbs the perfume emitted by the sample. Thereafter, the oil is extracted from the fat with alcohol and is then isolated later on. The success of such technique depends to a great extent upon the quality of the fat to be used. Paramount care must be exercised while preparing the corps for enfleurage. It must be practically odorless and of proper consistency. If the corps is too hard, the blossoms will not have sufficient contact surface with the fat, limiting its power of absorption and thus resulting in a subnormal yield. However, if the corps is too soft, it will tend to swallow up the flowers and the exhausted flowers will stick; when removed, the flowers will keep hold of the adhering fat, thus resulting in a substantial shrinkage and loss of corps. The consistency of the corps should be such that it offers a semihard surface from which the exhausted flowers can easily be removed.

6.1.2.6 Expression

For the production of citrus oils, techniques like expression or cold pressing is used. Expression refers to any physical process in which the glands of essential oil present within the peel are crushed or broken in order to release the oil. Spugna method was the method which was practiced many years ago in Sicily. The oil was removed from the peel either by pressing the peel against a hard object of baked clay called concolina which was placed under a large natural sponge or by bending the peel into the sponge. Later on the oil absorbed by the sponge was removed by squeezing. It is reported that oil produced from this technique contains more of the fruit odorous characteristics than oil produced by any other methods. Another method known was equaling technique which uses a shallow bowl of copper or brass with a hollow central tube; similar in shape to a shallow funnel. The bowl is equipped with brass points with blunt ends across which the whole fruit sample is rolled by hand with some pressure until all of the oil glands have burst. The oil and aqueous cell contents are allowed to drip down the hollow tube into a pot from which the oil is separated by decantation technique.

6.1.2.7 Infusion

Infusions are prepared by soaking a drug in water for a definite period of time. The process can be either hot or cold, depending upon the type of the ingredients present as decomposition may occur at higher temperatures. Infusions are generally prepared for immediate use, as preservatives are absent. In some cases preservatives like alcohol are used and the infusions concentrated by boiling. The term infusion are used for the preparations prepared from soft tissues like petals, leaves etc.

6.1.2.8 Decoction

Decoctions are prepared in a similar manner to that of infusions but the ingredients are boiled with that of water for a specified period of time or till a definite volume is attained. The term decoction is used when the preparation is prepared using hard plant parts like root, bark, wood etc. Decoction is usually the method of choice when working with tougher and more fibrous plants, barks and roots (and which have water soluble chemicals). Depending on the type of plant material being used, strong decoctions are prepared in two different ways. The first involves boiling the mixture for a longer duration. This is generally used when working with barks. Longer boiling time, up to 2 h or more, is sometimes necessary to break down, soften, and extract the larger pieces. On the other hand, when smaller woody materials are used the decoction is prepared as mentioned above and then it is allowed to soak overnight before straining out the herb.

6.2 REMOVAL OF UNWANTED AND INTERFERING COMPONENTS AFTER EXTRACTION

It has been seen that many naturally occurring compounds may hinder the isolation and purification of desired bioactives. A few general procedures are mentioned below that may help to realize that contamination may have taken place during extraction.

Lipids are usually extracted with low polar solvent and may be co-extracted when polar solvents are used. These compounds may be visualized on a TLC plate and using iodine vapor in a closed chamber to reveal brown spots or by performing an 1H-NMR :Proton Nuclear Magnetic Resonance spectral measurement in which a high broad peak around δ1.2–1.4 is generally observed.

- In order to remove the fats and waxes from an extract, the ground plant material may be percolated with low polar solvents (petroleum ether or hexanes) and allowed to dry prior to the full extraction process.
- Column chromatography or vacuum-liquid chromatography using nonpolar solvents as eluents may also be used to wash out the nonpolar components.
- Filtering sample through a reverse phase chromatographic column is another approach.
- An alternative approach is to add methanol or methanol-water in sufficient volume to dissolve the desired polar component, assisting the dissolution by ultrasound treatment if necessary and then filtering off the precipitate.

Unwanted pigments such as chlorophylls and flavonoids are also sometimes present at high concentration and depend on the plant part under use. Although they are not easily removed but the following methods may be applied.

- Use of Activated charcoal or activated carbon to decolorize solutions by a selective adsorption phenomenon.
- The solution may be either percolated through a relatively short charcoal column, or the powder can be mixed with the liquid to be decolorized, left to stand for a period of time, and filtered.

The efficiency of the adsorption is increased by heating. Charcoal has the disadvantage that many active medicinal compounds can also be adsorbed, as in the case of morphine, strychnine, and quinine.

- An aqueous solution of 2–5% lead-(II) acetate can be used for the removal of fatty acids, chlorophylls, and other colored materials. The gummy extract from the plant is dissolved in 95% aqueous ethanol, with warming if necessary, and the lead-(II) acetate solution is then added producing an insoluble precipitate, which is filtered off by suction through celite or diatomaceous earth.

Vegetable tannins are polyphenols commonly found in plant extracts, often in high concentrations, and give false positive results in many biological assays usually because of their tendency to precipitate proteins through multipoint hydrogen bonding. These compounds are extracted when using polar solvents, and it has been observed that aqueous and organic extracts containing tannins inhibit enzymes leading to false-positive results.

- The presence of tannins can be confirmed by the formation of a precipitate with ferric chloride, and they can also be removed from aqueous and nonpolar extracts by passage overpolyamide, collagen, Sephadex LII-20, or silica gel.

- Tannins may also be precipitated by gelatin-sodium chloride solution (5% w/v NaCl and 0.5% w/v gelatin), caffeine or protein. They are retained by soluble polyvinyl pyrrolidone or polyamide resins, by the formation of hydrogen bonds between the tannin phenolic hydroxyl groups and the amide units of the retaining agent.
- Phenolic compounds can also be removed by washing with 1% NaOH solution.

Plasticizers can be eliminated by distilling the solvents, filtering the sample through a reverse phase chromatographic column, or by filtering the extract or sample through porous alumina.

At present, there are a number of well-established methods available for extraction of bioactives from various sources. A suitable protocol for extraction can be designed only when the target analyte(s) and the overall aim have been determined. It is also useful to obtain as much information as possible on the physicochemical nature of the analyte(s) to be isolated. For unknown entities, at times it may be necessary to apply pilot extraction approach to find out the best possible methods. At the time of choosing and deciding upon a method, one should be open-minded enough to welcome and weigh the advantages and disadvantages of all available extraction methods, with special emphasis on their effectiveness and, obviously, the total cost involved. A continuous progress in the area of separation technology has increased the variety and variability of the extraction methods that can be successfully utilized in the extraction of potent bioactives. For any natural product researcher, it is therefore essential to become familiar with the newer approaches in the field of extraction technology.

FURTHER READING

Wang, L., Weller, C.L., 2006. Recent advances in extraction of nutraceuticals from plants. Trends in Food Science & Technology 17, 300–312.

Chapter 7

Innovative Extraction Process Design and Optimization Using Design of Experimental Approach

Chapter Outline

7.1 INTRODUCTION

Our "Phytotherapy Research Group" has been striving toward giving a new face to classical pharmacognosy research, and application of chemometrics is just the beginning. Our group is involved in the design and process optimization of microwave-assisted extraction (MAE) of botanicals with a

Essentials of Botanical Extraction. http://dx.doi.org/10.1016/B978-0-12-802325-9.00007-0
137

potential for scale-up for quite a few years with some significant publications (Mandal et al., 2008, 2009, 2010; Das et al. 2013). A comprehensive survey for the critical assessment of experimental design, which is a core area of study in chemometrics in the process optimization of extraction of botanicals, in particular, with special emphasis on modern extraction techniques has been presented here.

Roughly, experimental designs can be divided into screening designs and optimization designs. Experimental designs are applied for the optimization of various factors during the development of several analytical strategies such as Liquid–Liquid extraction, sample digestion procedures, and Chromatographic methods.

In this chapter, a brief explanation of the different aspects and methodologies related to modern extraction techniques of botanicals that have been subjected to experimental design will be reviewed in particular along with a general synopsis on different aspects of chemometrics, and the steps involved in its practice are presented. A detailed study on various factors and responses involved during the optimization is also presented. There has been an increase in the number of papers referenced by Scopus with search term "extraction", "optimization" and "response surface methodology" have seen an increased activity during 2004–2014 (Figure 7.1).

There have been numerous articles published on the application of experimental designs with newer extraction technologies of botanicals as depicted in Figures 7.2–7.4.

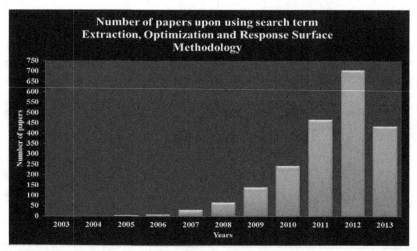

FIGURE 7.1 Number of papers referenced by Scopus since 2003 with search terms extraction, optimization, and response surface methodology. http://www.scopus.com/home.url, *Search conducted on December 28, 2013.*

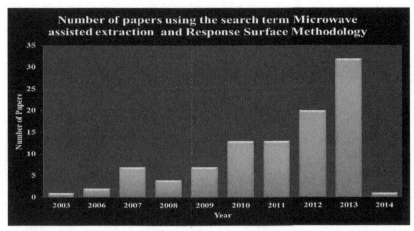

FIGURE 7.2   Number of papers referenced by Scopus since 2003 with search terms microwave assisted extraction (MAE) and response surface methodology. http://www.scopus.com/home.url, *Search conducted on December 28, 2013.*

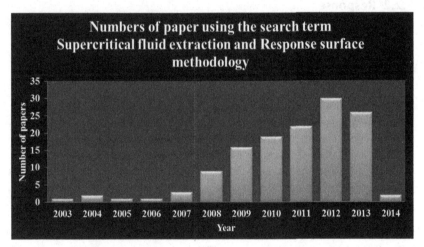

FIGURE 7.3   Number of papers referenced by Scopus since 2003 with search terms supercritical fluid extraction (SFE) and response surface methodology. http://www.scopus.com/home.url, *Search conducted on December 28, 2013.*

7.2 TERMINOLOGIES WE NEED TO KNOW

7.2.1 Experimental Design

This involves a statistical technique meant for planning, analyzing, conducting, and interpreting data obtained from preliminary experimental trials.

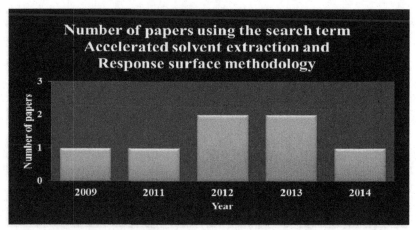

FIGURE 7.4 Number of papers referenced by Scopus since 2009 with search terms accelerated solvent extraction (ASE) and response surface methodology. http://www.scopus.com/home.url, *Search conducted on December 28, 2013.*

7.2.2 Response

This is the final outcome of an experiment that needs to be measured or observed, for example, yield of the targeted analyte and retention time.

7.2.3 Factor

This is a quantity that controls a response. Synonyms are parameter, parameter, independent parameter, and predictor. These are process inputs that an experimenter manipulates to induce a change in the output. They can be set and reset at different levels depending on the need and conditions affecting an experiment.

Let us suppose that an experimenter wishes to study the influence of six parameters or factors on the extraction of phytoconstituents using an industrial microwave. Figure 7.5 (a) and (b) illustrates the extraction process with possible inputs and outputs.

In the real-life scenario, some of the process parameters or the factors can be controlled very easily, whereas some of them are hard or expensive to control. In the diagram below, the output(s) help(s) to assess the performance of the entire process. The controllable factors can be varied during an experiment, and such factors actually dictate the overall process. Uncontrollable factors are difficult to control during the experiment, and such factors are responsible for the variability in product performance.

7.2.4 Level of a Factor

This denotes a value of a factor that is prescribed in an experimental design. Designs are named by the number of levels chosen for a factor, for example,

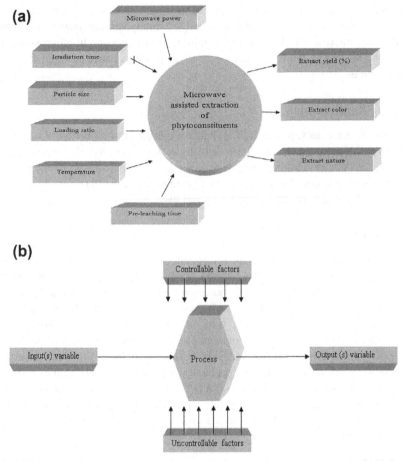

FIGURE 7.5 (a) Extraction process with possible inputs and outputs parameters. (b) Various controlled and uncontrolled factors or variables that can affect a system.

two-level, three-level designs. Examples are extraction temperature: 20, 25, 30 °C; extraction time: 1, 3, 5 min, ultrasound power: 20, 40, 60 kHz

7.2.5 Randomization

While designing and running an experiment, there are several factors in the form of external disturbances (commonly known as noise factors) that may actually influence the output of the experiment. For the above microwave extraction example, variations in the quality of the raw material due to seasonal change, variations in the temperature, humidity, and atmospheric pressure and their effects on the overall extraction yield, operator errors, power fluctuations may influence the final outcomes and such factors are hard to control.

Randomization is one of the methods to remove or reduce such errors occurring due to uncontrollable factors. Randomization actually helps in averaging out the effects of the external disturbances if present in the process. In other words, it can be said that all the levels of a factor have an equal chance of being affected by the noise factor.

7.2.6 Replication

Replication means repetitions of an entire experiment or a portion of it, under more than one operating condition. It helps to obtain an estimate of the experimental errors and to understand and estimate more specifically about the factors and their interaction. On the other hand, it requires time to conduct an experiment and replication may result in material loss.

7.2.7 Blocking

This is a mode to eliminate the effects of external disturbances and in the process actually improves the efficiency of the experimental design. External disturbances known to cause variation could be batch-to-batch variation, interday and intraday variations, shift-to-shift variation, etc. The primary aim is to arrange similar experimental runs into one group so that the whole group becomes a homogeneous unit. For example, an experimenter wants to improve the % yield of a drug through MAE. Four factors are being considered for the initial experimental trials that might have some impact on the extraction yield. It is decided to study each factor at a two-level setting (i.e., a low value and a high value). An eight experimental trial is chosen by the experimenter, but only four trials are possible to run per day. Here, each day can be treated as a separate block.

7.2.8 Response Surface

This is the relationship of a response to values of one or more factors. The surface is usually a plot in two or three dimensions of the function that is fitted to the experimental data. RSM is used to describe the use of experimental designs that give response surfaces from which information about the experimental system is deduced.

7.2.9 Model

This is an equation that relates a response to the factors/parameters under study. This means that the results or the outcomes can be described as a function based on the experimental factors, that is,

$$\mathbf{Y} = \mathbf{f}(x) + \varepsilon \text{ for one parameter } (x) \tag{7.1}$$

$$Y = f(x_1, x_2 \ldots x_n) + \varepsilon \text{ for two parameters } (x_1, x_2, x_3) \qquad (7.2)$$

The function $f(x)$ is represented by a polynomial function and denotes a good description of the relationships between the factors and the responses (Y) with residuals (ε). Three types of polynomial models are usually used in the design of the experiment and are hereby discussed in short with two parameters x_1, x_2.

- *Linear model*—This is the simplest polynomial model that contains only linear terms and describes only the linear relationships between the factors and the responses. A linear model with two parameters x_1, x_2 is expressed as

$$Y = b_0 + \sum_{i=1}^{k} b_i X_i + \varepsilon \text{ or } Y = b_0 + b_1 x_1 + b_2 x_2 + \varepsilon \qquad (7.3)$$

where Y is the estimated target function/response, b_i are the regression coefficients, b_0 is a constant (model intercept), i is the parameter/parameter number (1......k), and x_i is an independent parameter/parameter.

- *Interaction (second-order) model*—This is the next level of the polynomial model that contains additional terms that describe the interactions between different factors. It contains the following terms:

$$Y = b_0 + \sum_{i=1}^{k} b_i X_i + \sum_{i=1}^{k-1} \sum_{j=i+1}^{k} b_{ij} X_i X_j + \varepsilon$$

or

$$Y = b_0 + b_1 x_1 + b_2 x_2 + b_{12} x_1 x_2 + \varepsilon \qquad (7.4)$$

Here b_0, b_i, b_{ii}, and b_{ij} are the regression co-efficients for intercept, linear, quadratic, and interaction terms, respectively, and X_i, and X_j are the independent parameters/extraction parameters affecting the response (where $i \neq j$).

Both the linear and second-order models are mainly used for screening studies and robustness tests.

- *Quadratic model*—In order to determine the optimum value (maximum or minimum), quadratic terms need to be introduced in the model. By doing so, it is possible to determine nonlinear relationships between the factors and responses. This model contains the following terms:

$$Y = b_0 + \sum_{i=1}^{k} b_i X_i + \sum_{i=1}^{k} b_{ii} X_i^2 + \sum_{i=1}^{k-1} \sum_{j=i+1}^{k} b_{ij} X_i X_j + \varepsilon$$

or

$$Y = b_0 + b_1 x_1 + b_2 x_2 + b_{11} x_1^2 + b_{22} x_2^2 + b_{12} x_1 x_2 + \varepsilon \qquad (7.5)$$

7.2.10 Effects

This represents the coefficient of the factors in a model.

- *Main effect*—This denotes the coefficient of the terms in the first order of a factor.
- *Interaction effect*—This is the coefficient of the products of linear terms.
- *Quadratic effect*—This denotes the coefficient of the square of the linear terms.

7.3 ISSUES ADDRESSED THROUGH EXPERIMENTAL DESIGN

While developing a new methodology involving any type of extraction technique, two types of situations arise where the intervention of experimental design for their solution is necessary. The first is to screen out the "few vital " factors anticipated to have a significant effect on the final experimental outcome or response. The second is to find the selected factor's value that will optimize the response, that is, the optimization phase.

7.3.1 Screening Phase

In any experimental procedure, several experimental parameters or factors may influence the result. A screening experiment is performed in order to determine the experimental parameters and interactions that have a significant influence on the result, measured in one or several responses.

This screening process is schematically shown in Figure 7.6.

7.3.1.1 Some Advice before Initiating Screening Designs

- Specify the problem under study:
  - Review the whole procedure—different moments, critical steps, raw material, equipment, etc.
  - Try to get a general view of the problem.
- Define the responses:
  - Which response(s) can be measured?
  - Which source(s) of error(s) can be assumed?
  - Is it possible to follow the change in response(s) in due course of time?
- Selection of parameters:
  - Which experimental parameters are possible to study?
  - Review and evaluate the parameters—important, probably unimportant, etc.
  - Are all parameters interesting in the selected experimental domain?
  - Which interaction effects between the parameters can be expected?
  - Which parameters are probably not interacting and can be discarded?

This gives a list of possible responses or outcomes, experimental parameters, and potential interaction effects between the parameters. The time spent on planning in the beginning of a project is always paid back with interest at the end.

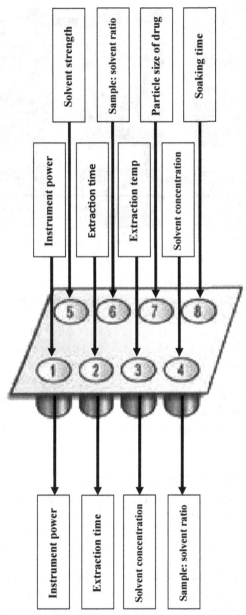

FIGURE 7.6 Schematic of a screening experimental design, where many factors are reduced to a significant few.

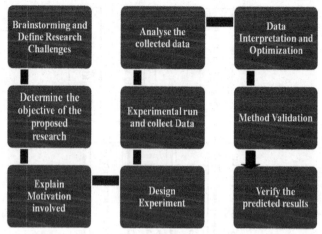

FIGURE 7.7   Work flow in an Experimental design approach.

- Hints on parameter selection

At this point, when the parameter to be investigated gets selected, it should also be decided as to which parameter need not be investigated. These unimportant parameters have to be kept at a fixed level in all experiments included in the experimental design. However, it must be remembered that it is always more economical to include a few extra parameters in the first screening than adding one parameter later. It is sometimes possible to lower the number of experiments needed, in order to achieve important information, just by redefining the original parameters.

An experimental design process actually follows a particular path as shown in Figure 7.7.

7.3.1.2 Screening Designs

When the list of factors gets finalized, an experimental design is chosen so as to estimate the influence of different factors on the final response. In screening studies, linear or second-order interaction models are common, such as full factorial or fractional factorial design because they are efficient and economical. The former is limited to the determination of the linear influence of the factors, that is, the main effects while the latter allows estimation of the interaction effects between the parameters in addition to the estimation of the main effects. Finally, the factors with the highest influence on the process are considered eventually.

7.3.1.2.1 Full Factorial or Factorial Design (2^k)

In a factorial design, the influence of all experimental factors and their interaction effects on the response(s) are investigated. If the combinations of k factors are investigated at two levels, a factorial design will consist of 2^k experiments. In Table 7.1, the factorial designs for 2, 3, and 4 experimental parameters are shown. To continue the example with higher numbers, six parameters would give $2^6 = 64$ experiments; seven parameters would render $2^7 = 128$ experiments, etc. The levels of the factors are given by (−) minus for low level and (+) plus for high level. A zero level is also included, a center, in which all parameters are set at their midvalue. Three or four center experiments should always be included in factorial designs, for the following reasons:

- the risk of missing nonlinear relationships in the middle of the intervals is minimized, and
- repetition allows for the determination of confidence intervals.

TABLE 7.1 Factorial Designs

2 factors			3 factors				4 factors				
Exp. No.	Factors		Exp. No.	Factors			Exp. No.	Factors			
	x_1	x_2		x_1	x_2	x_3		x_1	x_2	x_3	x_4
1	-	-	1	-	-	-	1	-	-	-	-
2	+	-	2	+	-	-	2	+	-	-	-
3	-	+	3	-	+	-	3	-	+	-	-
4	+	+	4	+	+	-	4	+	+	-	-
			5	-	-	+	5	-	-	+	-
			6	+	-	+	6	+	-	+	-
			7	-	+	+	7	-	+	+	-
			8	+	+	+	8	+	+	+	-
							9	-	-	-	+
							10	+	-	-	+
							11	-	+	-	+
							12	+	+	-	+
							13	-	-	+	+
							14	+	-	+	+
							15	-	+	+	+
							16	+	+	+	+

Note that all factors are changed simultaneously in a controlled way, to ensure that every experiment in each design is a unique combination of variable levels.

What – and + should correspond to for each parameter is defined from what is assumed to be a reasonable variation to investigate. In this way, the size of the experimental domain has been established. For two and three parameters, the experimental domain and design can be illustrated in a simple way. For two parameters, the experiments will describe the corners in a quadrate (Figure 7.8), while in a design with three parameters, they are the corners of a cube (Figure 7.9).

7.3.1.2.2 Fractional Factorial Design (2^{k-p})

During initial screening, the main aim is to obtain information on main effects and low order interactions. Thus, by fractionating the 2^k design (1/2, 1/4, 1/8, 1/16...1/2^p), factorial designs with 2^{k-p} experiments are required only where k is the number of factors and p is the size of the fraction. Thus, by fractionating

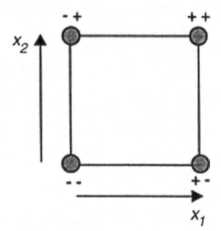

FIGURE 7.8 Experiments in a design with two parameters.

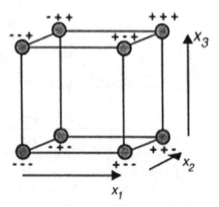

FIGURE 7.9 Experiments in a design with three parameters.

the 2^k design, the unwanted number of trials or runs can be removed. Moreover, fractionation leads to designs that give main effects only with a fewer runs. Hence, the factors that are identified as significant ones are then further investigated more deeply in the subsequent optimization phase.

7.3.1.2.3 Plackett–Burman Design (P-B)

A particular type of fractional factorial design is the P-B design, which assumes that the interactions can be completely ignored, and so the main effects are calculated with a reduced number of experiments. The P-B design is particularly popular for robustness tests in method validation studies. P-B designs allow one to examine the maximum of f = N−1 factors in N experiments, where N is a multiple of four (N = 8, 12, 16, 20...). The P-B design uses a first-order or linear model for screening, which is shown in

$$Y = b_0 + \sum_{i=1}^{k} b_i X_i + \varepsilon$$

(7.6)

7.3.2 Optimization Phase

The most important factors whether obtained from screening or from previous experiences are examined in detail to determine the optimal conditions using response surface designs for method development. In any extraction process, many extraction factors and interactions among them are known to affect the final outcome of the process or response. When such situations are anticipated, RSM is an effective tool for optimizing the process. Moreover, RSM has been successfully applied for various optimization procedures in extraction processes and pharmaceutical research. In RSM, an approximate relation between response and multiple factors is modeled as a polynomial equation obtained through regression analysis. The equation is called response surface and is represented graphically as a contour plot for analyzing the desired response or outcome.

7.3.2.1 Stages of Application of Response Surface Methodology

- **Stage 0**: At first, some ideas are generated concerning the factors or parameters that are likely to be important in response surface study. This is usually called a screening experiment. The objective of factor screening is to reduce the list of candidate parameters to a relatively few so that subsequent experiments will be more efficient and require fewer runs or tests. The purpose of this phase is the identification of the important independent parameters.
- **Stage 1**: Here, the objective is to determine the current settings of the factors, which will result in a response value close to the optimum region. If the current settings or levels of the independent parameters are not consistent with optimum performance, then the experimenter must determine a set of

adjustments to the process parameters that will move the process toward the optimum. This phase of RSM uses the first-order model without interaction, and the technique used is called the method of steepest ascent (descent).

- **Stage 2**: This generally begins when the process approaches near the optimum. Here, at this point, an experimenter wants a model that will accurately approximate the actual or true response function. As the true response surface usually exhibits a curvature near the optimum, a second-order model must be used. Once an appropriate approximating model has been obtained, this model may be analyzed to determine the optimal conditions for the process. This sequential experimental process is usually performed within some region of the independent parameter space called the operability region or experimentation region or the region of interest.

7.3.2.2 Optimization Designs

7.3.2.2.1 Central Composite Design

The central composite design (CCD) is also known as the Box–Wilson design, which contains an imbedded factorial or fractional factorial design (corners of a cube) with center points which is augmented with a group of "star points," which helps to calculate the curvature.

Thus, in examining "f" factors, a CCD requires the experiment number $N = 2^k + 2k + C_0$, where 2^k is a two-level full factorial design, $2k$ is a star design, and C_0 is the center point. For each factor, a CCD requires five levels, $-\alpha$, -1, 0, $+1$, and $+\alpha$.

7.3.2.2.2 Box–Behnken Design

Box–Behnken design (BBD) are a class of rotatable or nearly rotatable second-order designs based on three-level incomplete factorial designs. There are no axial points, and the design requires $2k (k-1) + C_0$ experiments in which there is only one center point. The use of BBD should be considered for experiments with greater than two factors and when it is anticipated that the optimum is known to lie in the middle of the factor ranges.

7.3.2.2.3 Doehlert Design

Unlike CCDs and BBDs, Doehlert designs are not rotatable, and the factors can be varied at a different number of levels. For a three-factor design type, one factor can be at three, one at five, and one at seven levels. It is a very useful and attractive alternative to the experimental design of the second order. The Doehlert matrix points correspond to the vertices of a hexagon generated from a regular simplex design, and generally the total number of data points in the planning equals $k^2 + k + pc$, where k is the number of factors and pc is the number of experiments at the center point. Further, this type of design requires a smaller number of experiments over the CCD and is therefore more efficient.

7.3.2.2.4 Three-Level Full Factorial Design

This design contains all possible combinations between the factors (k) and their levels L = 3, resulting in N = L^k = 3^k experiments, including one center point. Thus, in the case of two factors, nine experiments are needed, while for three factors, 27 experiments are needed.

7.4 RSM AS A TOOL FOR OPTIMIZATION IN MAE

The majority of the uses of the design of experiments in MAE can be classified as either method validation robustness studies or optimizing method conditions. Typically, two, three, or four factors are studied during the optimization phase. Table 7.2 lists out the reports of MAE using the experimental design for optimization. It is observed that the CCDs are most popular and widely used for optimization, even when BBD, Doehlert design or 3^3 full factorial design might be better alternatives.

TABLE 7.2 Report on the Use of Various Experimental Designs in the MAE of Botanicals

Method and Analyte	Experimental Design	Factors/Levels
I. Report on the Use of Experimental Design During Screening for Significant Factors in Microwave Assisted Extraction (MAE) Techniques		
1. Microwave-assisted Soxhlet extraction of total oil content from sunflower seeds	Full factorial design with 3 factors	Microwave power (25–75%) Extraction time (60–90 s) Extraction cycle number (10–20)
2. Focused microwave-assisted digestion of bean samples for nitrogen determination	Two-level full factorial design with 4 factors with 2 central points	Temperature of decomposition (280–320 °C) Volume of sulfuric acid (8–12 ml) Volume of hydrogen peroxide (6–10 ml) Mass of potassium sulfate (0.5–1.5 ml)
3. MAE of puerarin from *Radix puerariae*	Plackett-Burman design (PBD) with 8 factors and 1 dummy	Solvent: material ratio (20:1–30:1 ml/g) Mean particle size (0.15–0.25 mm) Ultrasound (without–with) Solvent type (ethanol–methanol) Solvent concentration (40–60%) Microwave power (100–300 W) Extraction time (40–80 s) Extraction cycle (1–3)

Continued

TABLE 7.2 Report on the Use of Various Experimental Designs in the MAE of Botanicals—Cont'd

Method and Analyte	Experimental Design	Factors/Levels
4. MAE extraction of polysaccharides from *Stigma maydis*	PBD with 6 factors	Temperature (30–100 °C) Liquid–solid ratio (20–100 ml/g) Microwave power (300–700 W) Time (5–30 min) Particle size (20–100) Origin of material (siping–nongan)
5. MAE of triterpenoid lupeol from *Ficus racemosa* leaves	PBD with 6 factors	Microwave power (20% W of 700–50 % W of 700) Irradiation time (1–3 min) Solvent: sample ratio (5:1–15:1 ml/g) Particle size (20–40 mesh) Methanol concentration (50 %v/v–100%v/v) Preleaching time (5–10 min)
6. MAE of antihepatotoxic triterpenoid from *Actinidia deliciosa* roots	Fractional factorial design with 4 factors	Ethanol concentration, (30–90%) Extraction time (10–50 min) Liquid–solid ratio (5–25 ml/g) Microwave power (200–600 W)
7. MAE of active components from Yuanhu Zhitong prescription	Fractional factorial design with 6 factors and 2 factors fixed at a constant value	Microwave power (500–850 W) Extraction time (10–40 min) Ethanolic level (40–80%) Particle size (10–40 mesh)
8. MAE of secoisolariciresinol diglucoside from flaxseed	Fractional factorial design with 4 factors	Microwave power (30–120 W) Extraction time (1–5 min) Molarity of NaOH (1–0.25 N) Power mode (30–60 s/min)
9. MAE of carbohydrates from corn starch	Fractional factorial design with 4 factors	Heating temperature (140–160 °C) Heating time (1–5 min) Solid–liquid ratio (1–0.5 g/20 ml) Come up time (2–4 min)
10. MAE of polycyclic aromatic hydrocarbons from wood samples	Two-level factorial design, with 3 factors	Sample mass (4–6 g) Temperature (90–120 °C) Extraction time (15–20 min)

TABLE 7.2 Report on the Use of Various Experimental Designs in the MAE of Botanicals—Cont'd

Method and Analyte	Experimental Design	Factors/Levels
11. MAE of phenolic content from edible *Cicerbita alpina* shoots	Full factorial design with 4 factors	Extraction temperature (40–90°C) Extraction time (5–25 min) Extraction solvent (50–100%) Sample quantity (1–3 g)
12. MAE of artemisinin from *Artemisia annua*	Fractional factorial design with 2 factors	Sample/solvent ratio (0.5/30–2/120 g/ml) Extraction time (0.5–2 min)

II. Report on the Use of Experimental Design along with Other Chemometric Tools in Conjugation or Separately in the Optimization of MAE Techniques

Method	Analyte	Designs (Factors/Level)
1. MAE from pomegranate peel	Phenolic compounds	Central composite design (3/2)
2. MAE from broccoli	Total phenolics	Central composite design (4/5)
3. MAE of milk thistle seeds	Silymarin	Central composite design (4/5)
4. MAE of ginseng	Saponin	Central composite design (2/5)
5. MAE of color pigments from various sample of Rubiaceae plants	Alizarin, Purpurin	Central composite design (4/2)
6. MAE of "sweet tea tree"	Polysaccharides	Central composite design (3/3)
7. MAE of Pistachio green hull	Phenolic compounds	Central composite design (3/3)
8. MAE of active components from Chinese quince	Total phenolic content	Central composite design (3/5)
9. MAE of *Vitis coignetiae*	Total polyphenol content	Central composite design (3/5)

Continued

TABLE 7.2 Report on the Use of Various Experimental Designs in the MAE of Botanicals—Cont'd

Method and Analyte	Experimental Design	Factors/Levels
10. MAE of lemon peel	Pectin	Central composite design (3/5)
11. MAE of corn starch	Carbohydrates	Central composite design (2/5)
12. MAE of rice grains	Melatonin	Central composite design (5/3)
13. MAE of *Camellia oleifera* fruit hull	Polyphenols	Central composite design (3/5)
14. MAE of *Tremella fungus*	Polysaccharides	Central composite design (3/3)
15. MAE of hematococcus algae	Astaxanthin– carotenoid	Central composite design (4/5)
16. MAE of flaxseed hull	Secoisolariciresinol diglucoside	Central composite design (3/3)
17. MAE of *A. deliciosa* root	Antihepatotoxic triterpenoid	Central composite design (2/3)
18. MAE of pigments from Rubiaceae plants	Alizarin, Purpurin	Central composite design (4/3)
19. MAE of *Portulaca oleracea* seed	Omega-3- polyunsaturated fatty acids	Central composite design (3/5)
20. MAE of *Radix astragali*	Flavonoid	Central composite design (4/5)
21. Isolation of flavonone glycoside by MAE from *Citrus unshin* fruit peels.	Hesperidin	Central composite design (2/5)

TABLE 7.2 Report on the Use of Various Experimental Designs in the MAE of Botanicals—Cont'd

Method and Analyte	Experimental Design	Factors/Levels
22. MAE of natural dye from *Bixa orellana* (Annatto) seeds	Dye	Central composite design (3/5) and artificial neural network
23. MAE of apple pomace	Pectin	Central composite design (4/5)
24. MAE of potato peels	Phenolic antioxidant	Central composite design (3/3)
25. MAE of *Ganoderma lucidum* fungus	Polysaccharides	Central composite design (4/3)
26. MAE of *Withania somnifera*	Withaferin A— steroidal lactone	Central composite design (4/5)
27. MAE of yellow horn	Seed oil	Central composite design (3/5)
28. MAE of grape seed	Phenolic antioxidants	Central composite design (3/3)
29. MAE of pomegranate peel	Total phenolic compounds	Central composite design (3/3)
30. MAE of fern	Antimicrobial oil	Central composite design (3/5)
31. MAE of *Vitis vinifera*	Terpenic compounds	Central composite design (3/5)
32. MAE of *Chaenomeles sinensis*	Oleanolic acid, Ursolic acid	Central composite design (3/5)
33. MAE of *Citrus mandarin* peels	Phenolic acids	Central composite design (4/5)
34. MAE of *Stigma maydis*	Carbohydrate polymer	Central composite design (3/3)
35. MAE of *R. puerariae*	Isoflavones—puerarin	Central composite design (4/5)

Continued

TABLE 7.2 Report on the Use of Various Experimental Designs in the MAE of Botanicals—Cont'd

Method and Analyte	Experimental Design	Factors/Levels
36. MAE of purple cabbage	Proanthocyanidins	Central composite design (4/2)
37. MAE of blackcurrant	Anthocyanins	Central composite design (3/5)
38. MAE of peanut hull	Dietary fibers	Central composite design (4/3)
39. MAE of onion skin	Quercetin	Central composite design (2/3)
40. Microwave-assisted ionic liquids extraction of *Fructus forsythia* seed	Essential oil	Central composite design (3/5)
41. MAE of brewer's spent grain	Polyphenols	Central composite design (4/2)
42. MAE of mango seed	Antioxidants	Central composite design (5/5)
43. MAE of pomegranate rind	Dye extraction	Central composite design (3/5) and artificial neural network
44. MAE of cherry laurel	Phenolic compounds	Three-level-three-factor full factorial design
45. MAE of sour cherry (*Prunus cerasus* var. Marasca)	Anthocyanins and phenolic acids	Central composite design
46. MAE of *Trigonellafoenum-graecum* L	Diosgenin	Central composite design (3/5)
47. MAE of *Taxus baccata*	Paclitaxel	2^4 Factorial design
48. MAE of *Basil* and *Epazote*	Essential oils	Two-level factorial design
49. MAE of wheat and corn	Zearalenone	Two-level factorial design

TABLE 7.2 Report on the Use of Various Experimental Designs in the MAE of Botanicals—Cont'd

Method and Analyte	Experimental Design	Factors/Levels
50. MAE of *curcuma longa*	Curcumin	Taguchi L_9 orthogonal design.
51. MAE of green coffee oil	Green coffee beans	Full factorial design
52. MAE of Catathelasma ventricosum fruiting	Polysaccharides	Box–Behnken design (3/3)
53. MAE of hawthorn fruit	Polyphenols	Box–Behnken design (3/3)
54. MAE of fungus	Polysaccharides	Box–Behnken design (3/3)
55. MAE of *Capsicum frutescens* L	Capsaicin	Box–Behnken design (3/3)
56. MAE of apple pomace	Polyphenols	Box–Behnken design (4/3)
57. MAE of *Platycoden grandiflorum* flower	Polysaccharides	Box–Behnken design (3/3)
58. MAE of *Salvia mirzayanii*	Essential oils	Box–Behnken design (3/3)
59. MAE of active components of Yuanhu zhitong prescription	Tetrahydropalmatine, Imperatorin, Isoimperatorin	Box–Behnken design (6/3)
60. MAE of bean sample	Nitrogen determination	Doehlert design (4/3)
61. MAE of industrial potato processing	Phenolic compound	Orthogonal array design
62. MAE of *F. racemosa* leaf	Triterpenoid—lupeol	Box–Behnken design (3/3)
63. MAE of sugar beet pulp	Pectin	Box–Behnken design (4/3)

Continued

TABLE 7.2 Report on the Use of Various Experimental Designs in the MAE of Botanicals—Cont'd

Method and Analyte	Experimental Design	Factors/Levels
64. MAE of *Perilla frutescens* leaves	Flavonoids	Box–Behnken design (3/3)
65. MAE of mulberry	Anthocyanins	Box–Behnken design (3/3)
66. MAE from waste peanut shells	Polyphenols	Box–Behnken design (4/3)
67. MAE of *Cryptotaenia japonica* Hassk	Flavonoids	Box–Behnken design (4/3)

7.5 RSM AS A TOOL FOR OPTIMIZATION IN SUPERCRITICAL FLUID EXTRACTION

Compared to other extraction techniques, the optimization of a supercritical fluid extraction (SFE) procedure is a complex process due to the multitude of parameters: extraction time, pressure, temperature, flow, tapping technique, and supercritical fluid composition. In addition to the large number of parameters, each factor can also have a marked effect on extraction efficiency. Therefore, the establishment of the optimal settings is both a very important and potentially time-consuming process. Several statistical techniques, such as factorial design and multilineal regression, have been employed in the optimization of analytical methods. Factorial design has some advantages in that the global optimum conditions can be provided, large amounts of quantitative information can be obtained, and both discrete and continuous factors can be estimated. Factorial designs have been used to determine the effect that numerous parameters have on the process, including temperature, pressure, pretreatment of sample, extraction time, fluid flow rate, and addition of a modifier. Table 7.3 lists out the reports of SFE using an experimental design for optimization.

7.6 RSM AS A TOOL FOR OPTIMIZATION IN PRESSURIZED LIQUID EXTRACTION/ACCELERATED SOLVENT EXTRACTION

One way of increasing the extraction rates of bioactive compounds from food matrices is by using pressurized liquid extraction (PLE). This technique uses high pressures, allowing the user to carry out extractions at temperatures above

TABLE 7.3 Report on the Use of Various Experimental Designs in the SFE of Botanicals

Method	Analyte	Designs (Factors/Level)
I. Report on the Use of the Experimental Design During Screening for Significant Factors in Supercritical Fluid Extraction (SFE) Techniques		
1. SFE of fatty acids from *Borago officinalis* L. Flower	Full factorial design	Temperature (35–65 °C) Pressure (atm) (100–350) Static extraction time (min) (10–40) Dynamic extraction time (min) (10–40) Modifier volume (ml) (0–100)
2. SFE of *Nigella sativa* seeds oil and its bioactive compound, thymoquinone	Taguchi design	Pressure (bar) (150–250) Temperature (°C) (40–60) CO_2 flow rate (g/min) (10–20)
II. Report on the use of Experimental Design in Optimization of SFE Techniques		
1. SFE of the flower of borage (*Borago officinalis* L.)	Fatty acids and essential oils	Central composite design
2. SFE of *Ginkgo biloba* L. leaves	Ginkgo biloba extraction yield	Box–Behnken design
3. SFE of *Artemisia annua* L.	Artemisinin	Central composite design
4. SFE of rapeseed (*Brassica napus* L.)	Rapeseed oil	Box–Behnken design
5. SFE of *Maydis stigma*	Flavonoids	Box–Behnken design
6. SFE of *Herba Moslae*	Essential oil	Box–Behnken design
7. SFE of *Spirulina platensis*	Antioxidant profiling	Box–Behnken design
8. SFE of *Asianpear*	Arbutin	Box–Behnken design
9. SFE of date pits	Lipids	Central composite design
10. SFE of *Lepidium apetalum*	Seed oil	Central composite design
11. SFE of apricot pomace	β-Carotene	Central composite design

Continued

TABLE 7.3 Report on the Use of Various Experimental Designs in the SFE of Botanicals—Cont'd

Method	Analyte	Designs (Factors/Level)
12. SFE of *Artemisia annua* L.	Scopoletin and Artemisinin	Central composite design
13. SFE of *Cydonia oblonga* Miller seeds	Fatty acids	Central composite design
14. SFE of *Tribulus terrestris*	Essential oil and Diosgenin	Central composite design
15. SFE of roselle seeds	Phytosterol	Central composite design
16. SFE of *Mentha spicata* L. Leaves	Bioactive flavonoid compounds	Central composite design
17. SFE of *Herba epimedii*	Icariin	Central composite design
18. SFE of *Spirulina platensis* ARM 740	γ-Linolenic acid	Central composite design
19. SFE of Passiflora seed	Oil	Central composite design
20. SFE of *Myrtus communis* L. leaves	Essential oils	Central composite design
21. SFE of wheat germ	Natural vitamin E	Central composite design
22. SFE of grape peel	Total phenols, antioxidants, and total anthocyanins	Orthogonal array design
23. SFE of *N. sativa* seeds	Thymoquinone	Full factorial design

the boiling point of the solvent. This enhances analyte solubility and mass transfer rates, resulting in better recoveries of the target compounds than in conventional solid–liquid extraction techniques. The high temperatures used in PLE also decrease the viscosity and the surface tension of the solvents. This means that the sample matrix area is accessed more easily, which also improves the extraction rate. In this way, PLE uses a combination of high pressures and temperatures that provide faster extraction processes that require small amounts of solvents. PLE also generally requires smaller amounts of organic solvents than do conventional techniques and could therefore be considered a green extraction technique.

TABLE 7.4 Report on the Use of Various Experimental Designs in ASE of Botanicals

Method	Analyte	Designs (Factors/Level)
I. Report on the Use of Experimental Design During Screening for Significant Factors in Pressurized Liquid Extraction (PLE)/accelerated Solvent Extraction (ASE) Technique		
1. ASE of olive leaves	Oleuropein content	Plackett–Burman design
2. ASE of Eiseniabicyclis (Kjellman) Setchell	Fucoxanthin	Plackett–Burman design
II. Report on the Use of Experimental Design in the Optimization PLE/ASE Technique		
1. ASE of jabuticaba skins	Phenolic compounds	Full factorial design
2. ASE of olive leaves	Oleuropein content	Central composite design
3. ASE of *Fructus Schisandrae*	Schizandrin, Schisandrol B, Deoxyschizandrin, and Schisandrin B	Box–Behnken design
4. ASE of rosemary (*Rosmarinus officinalis* L.), Marjoram (*Origanum majorana* L.) and Oregano (*Origanum vulgare* L.)	Antioxidant compounds	Central composite design
5. ASE of Spirulina platensis microalga	Antioxidant compounds	Full factorial (3 levels) design
6. ASE of *Piper gaudichaudianum* Kunth leaves	Nerolidol, palmitic acid, phytol, stearic acid, squalene, vitamin E, stigmasterol, and β-sitosterol	Full factorial design
7. ASE of *Cynanchum bungei*	4-Hydroxyacetophenone, baishouwubenzophenone, and 2,4-dihydroxyacetophenone	Box–Behnken design
8. ASE of potato peel	Caffeic acid content	Box–Behnken design
9. ASE of lemongrass (*Cymbopogon citratus*)	Oleoresin	Central composite design
10. ASE of sage (*Salvia officinalis* L.), basil (*Ocimum basilicum* L.), and thyme (*Thymus vulgaris* L.)	Phenolic antioxidants	Central composite design

Continued

TABLE 7.4 Report on the Use of Various Experimental Designs in ASE of Botanicals—Cont'd

Method	Analyte	Designs (Factors/Level)
11. ASE of *Angelica sinensis*	Z-ligustilide, Z-butylidenephthalide, and ferulic acid	Central composite design
12. ASE of *Orthosiphon stamineus*	Oil	Box–Behnken design
13. ASE of sweet potato	Anthocyanins	Central composite design
14. ASE of apple pomace	Antioxidants	Central composite design
15. ASE of small green cardamom (*Elettaria Cardamomum*) seed	Essential oil	Box–Behnken design

To date, PLE has mainly been used at the analytical level for the quantitative recovery of target analytes. For instance, PLE has been used to optimize the analytical extraction of polyphenols from apples using methanol as an organic solvent. The technique has also been used to optimize the sample preparation procedure for analyzing polyphenols from parsley. In recent times, the technique has shown potential as a green extraction method to obtain crude extracts with useful biological properties. For example, PLE was used to optimize the extraction of antioxidants from microalgae, anthocyanins from dried red grape skin, antioxidants from rosemary, and vitamin E rich oil was isolated from grape seeds using PLE. A number of variables can be optimized to maximize the yields of a target compound using PLE. Therefore, RSM could be a useful tool for optimizing extractions where a number of variables can be optimized (Table 7.4).

FURTHER READING

Das, A.K., Mandal, V., Mandal, S.C., 2013. Design of experiment approach for the process optimisation of microwave assisted extraction of lupeol from *Ficus racemosa* leaves using response surface methodology. Phytochem. Anal. 24, 230–247.

Mandal, V., Mohan, Y., Hemalatha, S., 2008. Microwave assisted extraction of curcumin by sample-solvent dual heating mechanism using Taguchi L9 orthogonal design. J. Pharm. Biomed. Anal. 46, 322–327.

Mandal, V., Dewanjee, S., Mandal, S.C., 2009. Microwave assisted extraction of total bioactive saponin fraction from *Gymnema sylvestre* with reference to gymnemagenin: a potential biomarker. Phytochem. Anal. 20, 491–497.

Mandal, V., Mandal, S.C., 2010. Design and performance evaluation of a microwave based low carbon yielding extraction technique for naturally occurring bioactive triterpenoid: oleanolic acid. Biochem. Eng. J. 50, 63–70.

Antony, J., 2003. Statistical Design and Analysis of Experiments with Applications to Engineering and Science. Elsevier. ISBN 0 7506 4709.

Chapter 8

Identification Strategies of Phytocompounds

Chapter Outline

8.1 IDENTIFICATION STRATEGY FOR VOLATILE COMPOUNDS

The identification strategy depends upon the availability of fluid ionization detector or mass spectra (MS) and Fourier transform infrared (FTIR) spectral data. Dependence on a single retention value, even when using the same capillary column, sometimes gives a false positive test. Kovat's retention index system offers an excellent means of gas chromatography (GC) identification. The following approaches can be adopted:

- Either the retention indices have to be identical (within a ±0.2 index unit) using three stationary phases with different selectivities (Figure 8.1(a)).
- When a single column is being used, the dependence of the retention index on temperature has to be identical at not less than five different temperatures (Figure 8.1(b)).

In both cases, identification can be carried out without reference compounds if reliable data can be obtained from the literature. Without a reference compound, the certainty of identification can be increased if, besides the MS spectra, retention indices on two different stationary phases should be used (Figure 8.1(c)). If only a single column is being used, FTIR should also be made available along with MS (Figure 8.1(d)).

Essentials of Botanical Extraction. http://dx.doi.org/10.1016/B978-0-12-802325-9.00008-2

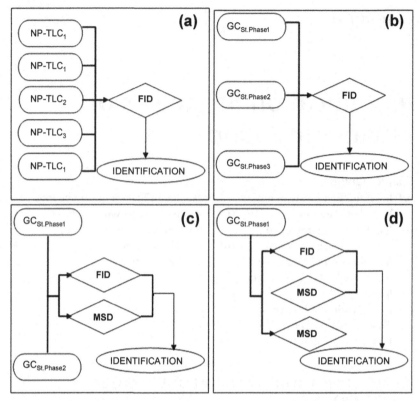

FIGURE 8.1 Identification strategies for volatile constituents. GC: Gas chromatography, St: Stationary phase, NP: Normal phase, TLC: Thin layer chromatography, FID: Flame ionization detector, MSD: Mass selective detector.

8.2 IDENTIFICATION STRATEGY OF NONVOLATILE COMPOUNDS

Three different approaches are made use of in the case of identification of non-volatile compounds.

1. For known compounds using a reference sample, three different separation methods and one spectroscopic method should be adopted.
2. For known compounds without a reference sample, two different spectro-scopic methods should be adopted.
3. For unknown compounds, one high-performance separation method and three different spectroscopic method should be adopted.

8.3 IDENTIFICATION OF KNOWN COMPOUNDS USING REFERENCE STANDARDS

The identification of phytocompounds by thin layer chromatography (TLC) is the basic technique given in several pharmacopoeias. TLC is performed

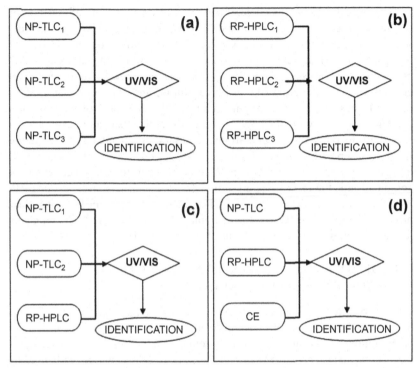

FIGURE 8.2 Identification strategies for known, nonvolatile constituents using reference standard strategies for volatile constituents. RP: Reverse phase.

under standard conditions, and the spot of the substance under study, with or without derivatization, is compared with that of a similarly developed reference material applied at approximately the same concentration. Using TLC for identification, the plant extract and the reference substances need to be always applied on the same plate and the homologous restriction factor (hR_f) values must be calculated from the densitograms. Using a single solvent system, two compounds can be considered as identical if the variance of the hR_F values of the compounds to be identified and the reference substances is less than ±3. However, if the hR_F value of the unknown compound has practically the same value as that of the reference substance in three different mobile phases with different selectivities, only then can it be considered identical with the reference substance. The identification scheme using three different mobile phases and in situ ultra violet (UV)–visible spectra is depicted in Figure 8.2(a). For the satisfactory identification of known components of plant extracts, not only the chromatographic but also the spectroscopic data must be identical. For in situ spectroscopic identification from TLC plates, two criteria must be fulfilled. First, every minimum and maximum of the UV and/or visible spectra must be practically identical and second the ratio of the local absorbance minima and maxima must be identical. Using reverse phase (RP) high-performance liquid chromatography (RP–HPLC) for known compounds and if reference compounds are present, generally three different mobile phase compositions (different solvent strength and selectivity) and diode array detector (DAD) need to be used for identification. The creation of significantly different mobile phase compositions is a real challenge, because only few solvents (methanol, ethanol, propanol, acetonitrile, tetrahydrofuran, and water) can be used for reversed phase separations (Figure 8.2(b)). The advantage of normal phase (NP) TLC (NP–TLC) is the large number of solvents that can be combined to vary the selectivity of the separations, while the disadvantage of the method is the low separation power. Using RP–HPLC, fewer solvents can be used; however, the separation power is much greater. Therefore, in practice, a good combination is the use of two different NP–TLC mobile phases and one RP–HPLC mobile phase (Figure 8.2(c)). For unconvincing identifications, three different separation methods, for example, NP–TLC, RP–HPLC, and Capillary electrophoresis (CE) must be used (Figure 8.2(d)).

8.4 IDENTIFICATION OF KNOWN COMPOUNDS WITHOUT REFERENCE STANDARDS

The use of retention data alone is insufficient for unambiguous identification because of the high risk of coelution of the compound in question with many other compounds in any separation system. Combined or coupled techniques, for example, HPLC–DAD–MS or HPLC–DAD–MS–MS are frequently used for structural elucidation and for identification of a known compound from a plant. The recent introduction of HPLC coupled to nuclear magnetic resonance (HPLC–NMR) presents a powerful complement to HPLC–DAD–MS screening. However, in many cases, two different separation methods, like RP–HPLC and NP–TLC (Figure 8.3(a)), or HPLC and CE (Figure 8.3(b)), or TLC and CE (Figure 8.3(c)) and the use of DAD and MS techniques are adequate for identification. Using TLC as one of the separation methods, UV/VIS can be carried out in situ on the chromatoplate, but MS can only be carried out offline. In practice, three types of identifications arise in the search for major compounds in medicinal plants:

- Identification of a known compound,
- Identification of an unknown compound,
- Verification of the presence of particular medicinal plants in an extract.

8.5 IDENTIFICATION OF COMPOUNDS WITH UNKNOWN STRUCTURES

The identification of compounds with unknown structures is a very difficult analytical task. Plant constituents of interest are usually isolated following a fractionation (mainly bioactivity-guided) procedure. In order to render this approach more efficient, the monitoring of plant extracts (crude or purified) with HPLC-hyphenated techniques avoids finding of known compounds and

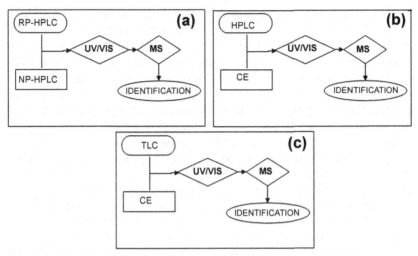

FIGURE 8.3 Identification strategies for known, nonvolatile constituents without a reference compound.

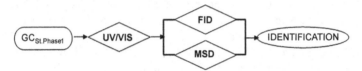

FIGURE 8.4 Identification strategies for unknown, nonvolatile constituents.

targets the isolation of new biologically active compounds. To provide a precise fingerprinting of the secondary metabolites in a given plant extract directly, online hyphenated techniques such as HPLC–DAD–MS–NMR are necessary as shown in Figure 8.4. This combination represents a valuable tool for further detailed metabolomic studies on either genetically modified plants or stressed plants. HPLC–DAD–MS–NMR allows the rapid structural determination of known compounds with only a minute amount of plant material. Simple bioautographic assays for screening biologically active compounds can also be performed directly online by collecting HPLC peaks and measuring activity. These bioassays permit a rapid location of the biologically active compounds. With such a combined approach, the time-consuming isolation of common natural substances is avoided, and an efficient targeted isolation of compounds with interesting biological and/or spectroscopic features can be performed.

8.6 THE STAGES IN STRUCTURAL ELUCIDATION

In examining the strategies for natural product structural determination, one generally has to go through four stages that are often overlapping in nature. First, there is the preliminary characterization in which the physical constants are determined, the molecular formula is established, and the functional groups are identified. At this stage, it is often possible to recognize the class of natural products to which the compound belongs. It may be apparent that the compound is known from another source or is just a simple derivative of another compound of a known structure. A simple chemical interrelationship may then be enough to establish the structure of the unknown compound. The second stage is structural simplification in order to identify the underlying carbon skeleton. Chemically, this work involves the selective removal of the various functional groups and the dissection of the carbon skeleton into identifiable fragments. Spectroscopically, it involves identifying adjacent groups of atoms. Evidence has to be accumulated concerning the position of the functional groups on the carbon skeleton. The determination of the position and relative stereochemistry of the functional groups forms the third phase. There is an interesting contrast in the use of spectroscopic and chemical methods between the first and the third stages of this strategy. In the first stage, methods are used to obtain information about the separate functional groups, but in the third phase, methods are used to reveal interactions between groups. The fourth stage involves establishing the absolute stereochemistry of the molecule. Although this is sometimes assumed, it is important to confirm it in order to understand the biological activity of a natural product. A few natural products are found in both enantiomeric forms. When a natural product or simple derivative crystallizes particularly well, X-ray crystallography can provide an unambiguous way of establishing the structure. When a structure has been proposed, it is helpful to rationalize this in biogenetic terms. An unambiguous partial synthesis from a compound of known structure, or a total synthesis, will finally provide the ultimate proof of structure.

The first stage in the structural determination is to establish the purity and to characterize the compound in terms of its elemental composition, empirical and molecular formulas. The criteria of purity and the basic analytical data that are used fall into four groups.

- Physical criteria: melting point or boiling point, optical rotation, refractive index.
- Analytical criteria: elemental composition determined by combustion analysis or high-resolution mass spectrometry, relative molecular mass.
- Chromatographic criteria: single spot on TLC, or single peak on GC or high-pressure liquid chromatography determined in several systems.
- Spectroscopic criteria: consistent relative integrals in the ^{1}H-NMR spectrum and consistent absorption in the infrared and UV spectrum.

However, the detailed description of all these techniques can be found in any standard analytical textbook.

The hyphenated technique of HPLC–DAD–MS–NMR allows structural determination of plant compounds from the raw extract, without isolation. Simple bioautographic assays for screening bioactive constituents can also be performed online by collecting the HPLC peaks and measuring the activity of interest. The advantages of this strategy are that compound/s will be isolated only if the structural elucidation establishes a novel constituent and/or the compound shows a high biological activity. The disadvantage of this strategy is that, for lack of isolated compounds, other bioactivity examinations cannot be carried out. Since the cost of such a combination of techniques is extremely high, only a few laboratories have these facilities at present. If the structure and/or bioactivity data generate interest, then further isolation strategy can be planned.

FURTHER READING

Nyiredy, S., December 5, 2004. Separation strategies of plant constituents—current status. Journal of Chromatography B 812 (1–2), 35–51.

Chapter 9

Qualitative Phytochemical Screening

Chapter Outline

9.1 DETECTION OF ALKALOIDS

The powdered drug is mixed thoroughly with 1 ml of 10% ammonia solution or 10% sodium carbonate solution, and then extracted by shaking for about 5 min with 5 ml methanol at 60 °C. The filtrate is cooled and then diluted enough for high-performance thin layer chromatography sample application.

1. Dragendroff's test
   To a few milliliters of the extract, add 2 ml of Dragendroff's reagent (potassium bismuth iodide). A reddish brown precipitate confirms the test as positive.
2. Mayer's test
   To a few milliliters of the extract, add a few drops of Mayer's reagent (potassiomercuric iodide). A cream colored precipitate indicates a positive test.
3. Wagner's test
   To a few milliliters of the extract, add a few drops of Wagner's reagent (solution of iodine in potassium iodide). A reddish brown precipitate indicates a positive test.

Essentials of Botanical Extraction. http://dx.doi.org/10.1016/B978-0-12-802325-9.00009-4

4. Hager's test

   To a few milliliters of the extract, add a few drops of Hager's reagent (saturated solution of picric acid). A yellow precipitate indicates a positive test.

5. Marme's test

   To a few milliliters of the extract, add Marme's reagent (cadmium iodide + potassium iodide + water). A precipitate is formed.

6. Scheibler's test

   To a few milliliters of the extract, add Scheibler's reagent (sodium tungstate + disodium phosphate + water). A precipitate is formed.

7. Reineckate test

   To a few milliters of the extract, add a few drops of reineckate solution (1 g of ammonium reineckate in water and 0.3 g of hydroxylamine hydrochloride in 100 ml ethanol). A precipitate is formed.

9.1.1 Thin Layer Chromatography Detection of Alkaloids

1. Without chemical treatment

   Ultraviolet (UV)-254 nm: Many alkaloids show a pronounced quenching of fluorescence in UV-254 nm.

   UV-365 nm: Some alkaloids (rauwolfia alkaloids, ajmaline) show intense blue or yellow fluorescence in UV-365 nm.

2. Spray reagents

   a. Dragendroff reagent: Brown or orange zones appear immediately on spraying. However, the color is not stable and can be made more distinct by spraying first with Dragendroff reagent and then with 5% sodium nitrite solution or 5% ethanolic sulfuric acid.

   b. Iodoplatinate reagent (IP): Directly after spraying, alkaloid zones appear brown, blue, or whitish on the blue–gray background of the thin layer chromatography (TLC) plate.

9.1.2 Solvent System

Solvent System	Remarks
Toluene–ethyl acetate–diethylamine (70:20:10)	Suitable for major alkaloids
Toluene–acetone–ethanol–concentrated ammonia (40:40:6:2)	Opium alkaloids
Chloroform–diethylamine (90:10)	Cinchona alkaloids
Acetone–water–concentrated ammonia (90:7:3)	Solanaceae drugs
Ethyl acetate–methanol–water (100:13.5:10)	Screening for rauwolfia alkaloids
Chloroform–methanol (85:15)	Isoquinoline alkaloids
Cyclohexane–chloroform–glacial acetic acid (45:45:10)	Berberine and protoberberine type alkaloids

9.1.3 Chemical Tests for Specific Alkaloids

- Tropane alkaloids
  Vitali–Morin's test: The extract is treated with fuming nitric acid and is evaporated to dryness on a water bath. The residue is then treated with 3% caustic potash. A deep purple color is formed.
  Rathenasinkam's test: In this test, nitric acid is used to effect the nitration of the benzene ring in atropine, hyoscyamine, and hyoscine. To the residue, add a few drops of ammonia and extract with chloroform. The chloroform extract is evaporated, and the residue is dissolved in acetone. A few drops of 10% sodium hydroxide is then added. A bluish-purple color is formed.

- Morphine
  Marquis' reagent: Formaldehyde in concentrated sulfuric acid. Purple red changing to purple is observed.
  Frshde's reagent: Molybdate in concentrated sulfuric acid. Violet, quickly changing to strong purplish red, fading out to weaker brown or brownish, then developing green. *Mecke's (or Lafon's) reagent* (Selenious acid in concentrated sulfuric acid). Green, quickly greenish blue, changing to blue, slowly to bluish green with a yellow–brown edge, then olivaceous green.

- Caffeine
  Muroxide test: In an evaporating dish, mix a couple of milligrams of the powdered substance, a minute crystal of potassium chlorate, and two drops of diluted hydrochloride (1: 1). Evaporate to dryness on a steam bath and continue heating several minutes longer. (If the residue remains completely colorless, heat cautiously with a small flame until it becomes pinkish or brownish.) Cool and treat with two drops of dilute ammonium hydroxide solution. A strong purple–red color is produced.

- Ipecacuanha alkaloids
  Frohde's test: This gives a greenish color with an alkaloidal solution.

- Cinchona alkaloids
  Thalleioquin test: To a dilute alkaloidal solution, add a few drops of bromine water. Shake well and then add a drop of strong ammonia solution. An emerald color is produced, which develops to a red color upon treatment with sulfuric acid.
  Erythroquinine test: To a dilute acidic solution of quinine, add a few drops of bromine water, a drop of 10% solution of potassium ferrocyanide, a drop of strong ammonia solution. A red color is formed.

- Indole alkaloids
  Van Urk's test or Elrich's test: To an alkaloidal solution, add a few drops of para dimethyl aminobenzaldehyde (Elrich's reagent), 5% ferric chloride and sulfuric acid. A purple color is formed.

- Nuxvomica
  Mandelin's test: To the sample solution, add Mandelin's reagent (sulfuric acid and ammonium vanadate). A violet blue color develops, which slowly changes to an orange color. The test is specific for strychnine.

Malaquin's test: To a dilute solution, add hydrochloride acid, zinc granules, and heat on a water bath at 100 °C for 5 min. Filter the solution and add a small crystal of sodium nitrate. A red color is formed.

9.2 DETECTION OF GLYCOSIDES

The powdered drug is extracted by heating for 15 min under reflux with 20 ml of 50% ethanol, with the addition of 10 ml of 10% lead acetate solution. After cooling and filtration, the clear solution is treated with a small quantity of acetic acid, then extracted by shaking with three 1 ml quantities of dichloromethane; shaking must be gentle to avoid emulsion formation. The combined lower phases are filtered over anhydrous sodium sulfate and finally evaporated to dryness. The residue is dissolved in 1 ml dichloromethane–ethanol (1:1), and this solution is used for TLC analysis.

1. Keller kiliani test: this test is specific for digitoxose moiety.
2. Baljet test: This comprises sodium picrate solution. A yellow color is observed which changes to orange.
3. Raymond test: Cardenolides on reaction with *m*-dinitrobenzene and methanolic potassium hydroxide give a purple color.
4. Legal's test: This comprises an alkaline sodium nitroprusside solution. A pink color is obtained.
5. Libermann–Burchard test: To a solution of the drug in glacial acetic acid, a drop of concentrated sulfuric acid is added. A change of color occurs from rose, through red, violet, and blue to green. The reaction is due to the steroid moiety.

Cardeinolides give a positive reaction for all the tests mentioned above. But bufadeinolides give a positive test for the Libermann–Burchard test only.

6. Borntrager's test: Boil the powdered drug with dilute sulfuric acid, filter and add chloroform to the filtrate. Shake well and collect the organic layer. Add a few drops of strong ammonia solution, shake slightly, and keep the test tube aside for few minutes. The lower ammoniacal layer takes on a pink or red color. This test is due to the presence of anthraquinones and is negative in the case of reduced forms of anthraquinones, that is, anthranols.
7. Modified Borntrager'test: This employs ferric chloride with dilute hydrochloric acid to bring about oxidative hydrolysis. The anthraquinones liberated are extracted with carbon tetrachloride and give a rose–pink to cherry red color when their solution is shaken with dilute ammonia.

9.2.1 TLC Detection

1. Without chemical treatment
   Fluorescence quenching by cardenolides is only very weak in UV-265 nm, but more distinct zones of fluorescence quenching are produced by bufadenolides. Cardiac glycosides do not fluoresce in UV-365 nm.

2. Spray reagents

   a. Specific detection of the five-member lactone ring (cardenolides)

      – Kedde reagent: Immediately on spraying, cardenolides form a pink or blue–violet color. Bufadienolides do not respond to this test.

      – Legal reagent, Baljet reagent, Raymond reagent (alkaline meta dinitrobenzene) gives a red–orange or violet color with cardenolides.

   b. General detection methods for cardenolides and bufadienolides

      – Chloramine–trichloroacetic acid reagent (CTA): Blue, yellow, or yellow–green fluorescent zones are observed in UV-365 nm.

      – Sulfuric acid reagent: The TLC plate is sprayed with 5 ml of reagent, then heated for 3–5 min at 100 °C. Blue, brown, green, and yellowish fluorescent zones are seen in UV-365 nm; the same zones appear brown or blue in daylight.

9.3 DETECTION OF FLAVONOIDS

The powdered drug is extracted with 10 ml methanol for 5 min on a water bath at about 60 °C. The clear filtrate is used for TLC analysis.

1. Shinoda test: To the plant extract, add a mixture containing a piece of magnesium ribbon and concentrated hydrochloric acid. The formation of a red color indicates flavonoids, flavonones, and xanthone.

2. To the test solution, add ferric chloride solution. A change of color from green to black occurs.

9.3.1 TLC Detection

1. Without chemical treatment

   UV-254 nm: All flavonoids cause fluorescence quenching, which is seen as dark blue zones on the yellow background of the TLC plate.

   UV-365 nm: Depending on the structure, flavonoids can give yellow, blue, or green fluorescence. Flavonoid extracts often contain other materials, such as plant acids and coumarins, which also show blue fluorescent areas.

2. Spray reagents

   a. Natural product reagent/polyethylene glycol (NP/PEG): Typical intense fluorescent colors in UV-365 nm are produced immediately on spraying, or after about 15 min. The addition of PEG lowers the detection limit.

   b. Fast blue salt reagent (FBS): Blue or blue–violet azo dyes are formed in daylight. To some extent, these can be intensified by further spraying with 0.1 M sodium hydroxide or 10% potassium hydroxide

9.4 DETECTION OF COUMARIN DRUGS

The powdered drug is extracted by shaking with 10 ml methanol for 30 min on a water bath. The clear filtrate is evaporated to about 1 ml and used for TLC analysis. The active principles in coumarin drugs are benzo-α-pyrones.

9.4.1 TLC Detection

1. Without chemical treatment

   UV-254 nm: All coumarins shows a distinct fluorescence quenching.

   UV-365 nm: All simple coumarins show an intense blue or blue–green fluorescence. Nonsubstituted coumarin fluoresces yellow–green in UV-365 nm only after treatment with potassium hydroxide.

2. Spray reagents

   a. Potassium hydroxide: Blue fluorescent zones are intensified by spraying with 5% ethanolic potassium hydroxide.

   b. NP/PEG: This reagent intensifies and stabilizes the existing fluorescence of the native coumarins.

   c. Antimony III chloride: Visnagin gives a lemon yellow fluorescence in UV-365 nm after treatment with this reagent.

9.5 DETECTION OF ESSENTIAL OILS

Steam distillation, extraction with dichloromethane, or extraction with methanol can be used as the sample preparation method for detection of essential oils.

1. Sample treated with tincture of sudan III gives a violet color.
2. Halphen's test: This test is used for the detection of cotton seed oil, a frequent adulterant. It is also known as Bevan's test. Two milliliters of the oil is mixed with 1 ml of amyl alcohol and 1 ml of 1% solution of sulfur in carbon disulfide for 10 min in a water bath. The formation of a red color indicates the presence of cotton seed oil.
3. Boudouin's test: This test is used to detect sesame oil. The basis of the test is the characteristic phenolic component sesamol, present in the oil. The oil is shaken with half its volume of concentrated hydrochloric acid containing 1% of sucrose. The development of pink color indicates the presence of sesamol.

9.5.1 TLC Detection

1. Without chemical treatment

   UV-254 nm: All compounds containing at least two conjugated double bonds quench fluorescence and appear as dark zones against the light green fluorescent background of the TLC plate. Phenylpropane derivatives exhibit this property, for example, anethole, eugenol, and myristicin. Other compounds that quench fluorescence are cinnamic aldehyde, thymol, and piperitone.

   UV-365 nm: An intense blue fluorescence is produced.

2. Spray reagents

   a. Anisaldehyde–sulfuric acid: In the visible range of light, the components of essential oils show strong blue, green, red, and brown coloration. Some compounds also show fluorescence under UV-365 nm.

  b. Vanillin–sulfuric acid: In the visible range, the observation is similar to that produced by anisaldehyde–sulfuric acid reagent.
  c. Phosphomolybdic acid: Constituents of essential oils show a uniform blue coloration on a yellow background when viewed in visible light (fenchone and anisaldehyde being the exception).

9.6 DETECTION OF CARBOHYDRATES

The following are some of the useful tests for sugars and other carbohydrates.

1. Fehling's test: To the heated solution of the substance, slowly add a mixture of equal parts of Fehling's solution No. 1 and No. 2. In certain cases, reduction takes place near the boiling point and is shown by a brick red precipitate of cuprous oxide. Reducing sugars include all monosaccharides, and some disaccharides (lactose, maltose, cellobiose, and gentiobiose). Nonreducing sugars include some disaccharides (sucrose) and polysaccharides.
2. Molisch's test: All carbohydrates give a purple color when treated with α-naphthol and concentrated sulfuric acid. With a soluble carbohydrate, this appears as a ring if the sulfuric acid is gently poured in to form a layer below the aqueous solution. However, in the case of an insoluble carbohydrate (cellulose), the color will develop only on shaking.
3. Osazone formation: Osazones are sugar derivatives formed by heating a sugar solution with phenylhydrazine hydrochloride, sodium acetate, and acetic acid. The yellow crystals formed can be examined under a microscope. Sucrose does not form osazone, but under the test conditions described, sufficient hydrolysis takes place for the formation of glucosazone.
4. Resorcinol test for ketones: This is known also as Selivanoff's test. A crystal of resorcinol is added to the solution and warmed on a water bath with an equal volume of concentrated hydrochloric acid. A rose red color indicates the presence of ketone (fructose, honey).
5. Test for pentoses: The test solution is heated in a test tube with an equal volume of hydrochloric acid containing a little phloroglucinol. The formation of a red color indicates the presence of pentoses.
6. Keller–killani test for deoxy sugars: Deoxy sugars are found in cardiac glycosides like Digitalis and Strophanthus. The sugar is dissolved in acetic acid containing a trace of ferric chloride and transferred to a test tube containing concentrated sulfuric acid. At the junction of the liquids, a reddish brown color is produced, which slowly turns blue.
7. Barfoed's test: When sugars are heated with Barfoed's reagent for 2 min, a red color is produced.

9.7 DETECTION OF PROTEINS AND AMINO ACIDS

1. Millon's test: To the aqueous extract, add a few drops of Millon's reagent. A white precipitate formed indicates the presence of proteins.

2. Biuret test: To an aliquot of the aqueous extract, add a few drops of 2% copper sulfate solution. To this, add 1 ml ethanol (95%), followed by excess potassium hydroxide solution. A pink color in the ethanolic layer indicates the presence of proteins.
3. Ninhydrin test: Two drops of ninhydrin solution are added to the aqueous extract. A characteristics purple color indicates the presence of amino acids.

9.8 DETECTION OF TRITERPENOIDS

1. Noller's test: To the test solution, add Noller's reagent (0.1% stannic chloride in thionyl chloride). A red color is produced.
2. Sannie test: A mixture of stannous chloride, acetic acid, and carbon tetrachloride (6:50:50) when sprayed on a filter paper containing a spot of triterpenes and heated at 100 °C produces a brown color.
3. Rosenthaler test: Addition of sulfuric acid to an alcoholic solution of triterpenes containing vanillin hydrochloride gives a color reaction.

9.9 DETECTION OF STEROIDS

1. Libermann–Burchard test: As described in Section 6.2.
2. Lifschutz reaction: A color reaction is produced when a small quantity of sterol is heated with perbenzoic acid, glacial acetic acid, and sulfuric acid.
3. Rossenhein reaction: The compound is treated with chloroform and sprayed with trichloroacetic acid. A Rossenhein color is produced for ergosterol.
4. Zimmermann test: This test is positive for all 17-keto sterols. The compound is treated with 1 ml of 2 N potassium hydroxide in absolute alcohol and 1 ml of 1% dinitrobenzene in absolute alcohol. After 10 min, the mixture was added to 10 ml of absolute alcohol. A violet color is produced.
5. Tschugaeff reaction: Glacial acetic acid solution of a sterol is treated with zinc chloride and acetyl chloride and boiled. A red color is produced.
6. Pinus reaction: Androsterone added to a solution of antimony trichloride in acetic acid gives a blue color.
7. Pettenkofer reaction: A solution of furfural in acetic acid added to dehydroepiandrosterone followed by the addition of sulfuric acid and warming gives a red color. This color changes to bluish red in several days.

9.10 DETECTION OF TANNINS AND PHENOLIC COMPOUNDS

1. Ferric chloride test: To the aqueous extract solution, add few drops of neutral 5% ferric chloride. A dark green color indicates the presence of phenolic compounds.

2. Gelatin test: To the aqueous extract, add 2 ml of 1% solution of gelatin containing 10% sodium chloride. A white precipitate indicates the presence of phenolic compounds. Gallic acid and other pseudotannins also precipitate gelatin if the solutions are sufficiently concentrated.
3. Goldbeater's skin test: Soak a small portion of Goldbeater's skin in 2% hydrochloric acid; rinse with distilled water and place in the test solution for 5 min. Wash with distilled water and transfer to a 1% solution of ferrous sulfate. A brown or black color on the skin denotes the presence of tannins. Goldbeater's skin is a membrane prepared from the intestine of the ox and behaves similarly to an untanned hide.
4. Phenazone test: To about 5 ml of an aqueous extract of the drug, add 0.5 g of sodium acid phosphate; warm, cool, and filter. To the filtrate add 2% solution of phenazone. All tannins are precipitated, the precipitate being bulky and often colored.
5. Test for catechins: Catechins on heating with acids form phloroglucinol, and they can, therefore, be detected by a modification of the well-known test for lignin. Dip a matchstick into the plant extract, dry, moisten with concentrated hydrochloric acid, and warm near a flame. The phloroglucinol produced turns the wood pink or red.
6. Test for chlorogenic acid: An extract containing chlorogenic acid when treated with aqueous ammonia and exposed to air gradually develops a green color.
7. Lead acetate test: The aqueous extract is dissolved in distilled water, and to this is added 3 ml of 10% lead acetate solution. A bulky white precipitate indicates the presence of phenolic compounds.

9.11 SPRAY REAGENTS

1. Anisaldehyde–acetic acid reagent: Anisaldehyde (0.5 ml) is mixed with 10 ml of 98% acetic acid. The TLC plate is sprayed with 5–10 ml, and then heated at 120 °C for 7–10 min in a drying cabinet. The plate may be sprayed afterward with concentrated sulfuric acid and evaluated in visible or UV-365 nm.
2. Anisaldehyde–sulfuric acid reagent: Anisaldehyde (0.5 ml) is mixed with 10 ml glacial acetic acid, followed by 85 ml methanol and 5 ml concentrated sulfuric acid. The TLC plate is sprayed with about 10 ml, heated at 100 °C for 5–10 min, and then evaluated in visible or UV-365 nm. This is useful in the detection of essential oils, bitter principles, pungent principles, and saponins.
3. Antimony (III) chloride reagent: A solution of antimony (III) chloride (20%) in chloroform. The TLC plate should be sprayed with 15–20 ml of the reagent, then heated for 5–6 min at 100 °C. Evaluation can be done in visible or UV-365 nm. This is useful for the detection of cardiac glycosides, saponins, visnagin, etc.

4. Benzidine reagent: Benzidine (0.5 g) is dissolved in 10 ml glacial acetic acid, and the volume is adjusted to 100 ml with ethanol. Evaluation can be done in visible light.

5. Berlin blue reagent: A freshly prepared solution of 10 g of iron(III) chloride and 0.5 g of potassium hexacyanoferrate in 100 ml of water. The plate is sprayed with 5–8 ml and evaluated in visible light. This is useful in the detection of arbutin.

6. Blood reagent: Ten milliliters of 3.6% sodium citrate is added to 90 ml fresh bovine blood. Two milliliters of this mixture is mixed with 30 ml phosphate buffer of pH 7.4. The plate must be sprayed in a horizontal position. This is useful for the detection of saponins, where white zones are formed against the reddish background of the plate. In some cases, warming of the plate may be required to effect hemolysis.

7. CTA: Ten milliliters of freshly prepared 3% aqueous chloramines T solution (sodium sulfamide chloride) is mixed with 40 ml of 25% ethanolic trichloroacetic acid. The plate is sprayed with 10–15 ml of the reagent and then heated at 100 °C for 5–10 min. Evaluation can be done in UV-365 nm. This is useful for the detection of cardiac glycosides.

8. Dichloroquinonechloramide reagent: This comprises 1% methanolic solution of 2.6-dichloroquinonechloroimide. The plate is to be sprayed with 5–10 ml of the reagent, and should be immediately exposed to ammonia vapors. This is useful in the detection of arbutin and capsaicin.

9. Dinitrophenylhydrazine reagent: 2,4-Dinitrophenylhydrazine (0.1 g) is dissolved in 100 ml methanol, followed by the addition of 1 ml of 36% hydrochloric acid. The plate should be immediately evaluated in visible light. This is useful in the detection of chromogenic dienes (*Valerianae radix*).

10. Dragendroff reagents : This is useful for the detection of alkaloids, heterocyclic nitrogen compounds, and quaternary amines.
    a. Dragendroff reagent: Basic bismuth nitrate (0.85 g) is dissolved in 40 ml water and 10 ml glacial acetic acid, followed by the addition of 8 g of potassium iodide dissolved in 20 ml of water.
    b. Dragendroff reagent R1: Tartaric acid (100 g) is dissolved in 400 ml of water. Basic bismuth nitrate (8.5 g) is added and the solution is shaken for 2 h. Two hundred milliliters of 40% potassium iodide is then added, and the solution is shaken vigorously. After standing for 24 h, the solution is filtered.
    c. Dragendroff reagent with tartaric acid: Solution A—17 g of bismuth subnitrate and 200 g of tartaric acid in 800 ml water. Solution B—160 g of potassium iodide in 400 ml water. Stock solution—A + B. Spray reagent—50 ml stock solution + 500 ml water + 100 g tartaric acid.
    d. Dragendroff reagent with hydrochloric acid (modified): Solution A—0.3 g of bismuth subnitrate, 1 ml of 25% hydrochloric acid and 5 ml water. Solution B—3 g of potassium iodide in 5 ml water. Spray reagent—5 ml of A + 5 ml of 12.5% HCl + 5 ml of B + 100 ml water.

e. Dragendroff reagent, followed by sodium nitrite: After treatment with any type of Dragendroff reagent, the plate may be additionally sprayed with 5% aqueous sodium nitrite or with 5% ethanolic sulfuric acid, for intensifying the colored zones.

11. FBS: Fast blue salt B (0.5 g; 3,3′-dimethoxy-biphenyl-4-4′-bis diazonium dichloride) is dissolved in 100 ml water. The plate is sprayed with 6–8 ml of the reagent, dried, and inspected in visible light. This is useful for the detection of bitter principles from hops and phenolic compounds.

12. Fast red salt reagent: Aqueous solution (0.5 g) of fast red salt B (diazotized 5-nitro-2-aminoanisole). The plate is sprayed with 10 ml, followed immediately by either 0.1 M sodium hydroxide or exposure to ammonia vapors. This is useful in the detection of amarogentin.

13. Ferric chloride reagent: This comprises of a 10% aqueous solution. Evaluation after spraying to be done in visible light. This is useful for the detection of oleuropein and hop bitter principles.

14. EP reagent: 4-Dimethyl aminobenzaldehyde (0.25 g) is dissolved in a mixture of 45 ml of 98% acetic acid. Five milliliters of 85% o-phosphoric acid and 45 ml water, followed by 50 ml concentrated sulfuric acid with cooling. The plate is then evaluated in visible light. The natural blue of azulenes is intensified using EP reagent.

15. Libermann–Burchard reagent: Acetic anhydride (5 ml) and 5 ml concentrated sulfuric acid are added carefully to 50 ml absolute ethanol, while cooling in ice. The reagent must be freshly prepared. The sprayed plate is warmed at 100 °C for 5–10 min, and then inspected in UV-365 nm. This is useful for the detection of triterpenes and steroids.

16. Iodine–chloroform reagent (I/CHCl₃): This comprises 0.5% iodine in chloroform. The plate after spraying is warmed at 60 °C for about 5 min and it may be evaluated at room temperature. This is useful for the detection of Ipecacuanha alkaloids.

17. Iodine reagent: Ten grams of solid iodine is sprayed on the bottom of a chromatography tank. Compounds possessing conjugated double bonds give yellow–brown zones in visible light, on exposure to the atmosphere of iodine vapors.

18. IP: Hydrogen hexachloroplatinate (IV) hydrate (0.3 g) is dissolved in 100 ml water and mixed with 100 ml of 6% potassium iodide solution. Evaluation of the plate is to be done in visible light. This is useful for the detection of nitrogen-containing compounds like alkaloids.

19. Potassium hydroxide: Ethanolic potassium hydroxide (5% or 10%; Borntrager reagent). The plate is sprayed with 10 ml and evaluated in visible light or in UV-365 nm, with or without warming. This is useful in the detection of anthraquinones (red), anthrones (yellow, UV-365 nm); coumarins (blue, UV-365 nm).

20. Kedde reagent (Kedde): Five milliliters of freshly prepared 3% ethanolic 3,5-dinitrobenzoic acid is mixed with 5 ml of 2 M sodium hydroxide. The plate is sprayed with 5–8 ml of the freshly prepared mixture and evaluated in visible light. This is useful for the detection of cardenolides.

21. Komarowsky reagent: One milliliter of 50% ethanolic sulfuric acid and 10 ml of 2% methanolic 4-hydroxybenzaldehyde are mixed just before use. The sprayed plate is heated at 100 °C for 5–10 min and evaluated in visible light. This is useful in the detection of essential oils, pungent principles, bitter principles, saponins.

22. Marquis reagent: Formaldehyde (3 ml) is diluted to 100 ml with concentrated sulfuric acid. The plate is evaluated in visible light immediately after spraying. This is useful in the detection of morphine, codeine, and thebaine.

23. Millon's reagent: Mercury (3 ml) is dissolved in 27 ml of fuming nitric acid, and the solution is diluted with an equal volume of water. This is useful for the detection of phenol glycosides in general.

24. Phosphomolybdic acid reagent: This comprises 20% ethanolic solution of phosphomolybdic acid. The plate is sprayed with 10 ml of the reagent and then heated at 100 °C for 5 min. This is useful in the detection of essential oils.

25. NP/PEG: The plate is sprayed with 1% methanolic diphenylboric acid-β-ethylamino ester (NP), followed by 5% ethanolic PEG-4000. Intense fluorescence is produced immediately or sometimes after 15 min in UV-365 nm. This is useful for the detection of flavonoids, aloin.

26. Ninhydrin reagent: Ninhydrin (30 g) is dissolved in 10 ml n-butanol, followed by 0.3 ml of 98% acetic acid. After spraying, the plate is heated for 5–10 min and then inspected under visible light. This is useful for the detection of amino acids and biogenic amines.

27. Nitrosodimethylaniline reagent: Nitrosodimethylaniline (10 mg) is dissolved in 10 ml pyridine, and used immediately for spraying. Gray–blue zones appear in visible light. This is useful for the detection of anthrone derivatives.

28. Sulfuric acid: Ethanolic sulfuric acid (5% or 10%), 50% ethanolic sulfuric acid, concentrated sulfuric acid are generally used as spraying reagents. After spraying, the plate is heated at 100 °C for 5 min and then evaluated in visible light or in UV-365 nm. Colored zones appear immediately. This is useful in the detection of cardiac glycosides.

29. Vanillin–phosphoric acid reagent: A—1 g of vanillin dissolved in 100 ml of 50% phosphoric acid. B—two parts of 24% phosphoric acid and eight parts of 2% ethanolic vanillic acid. After spraying with either A or B, the plate is heated for 10 min at 100 °C, and evaluated in visible or under Uv-365 nm. This is useful in the detection of lignanes, terpenoids.

30. Vanillin–hydrochloric acid reagent (VHCl): The plate is sprayed with 5 ml of 1% ethanolic vanillin, followed by 3 ml concentrated hydrochloric acid, then evaluated in visible light. Colors can be intensified by heating for 5 min at 100 °C.

31. Vanillin-sulfuric acid reagent: Solution I—5% ethanolic sulfuric acid, Solution II—1% ethanolic vanillin. The plate is sprayed with 10 ml of solution I, followed by immediate spraying of 5–10 ml of solution II. After heating at 110 °C for 5–10 min, the plate is evaluated under visible light. This is useful in the detection of essential oils.

32. Van URK reagent: 4-Dimethyl aminobenzaldehyde (0.2 g) is dissolved in 100 ml of 25% hydrochloric acid with the addition of one drop of 10% ferric chloride solution. This is useful in the detection of indole alkaloids.

FURTHER READING

Houghton, P., Raman, A., 1998. Laboratory Handbook for the Fractionation of Natural Extracts. Chapman & Hall, London, UK.

Chapter 10

Profiling Crude Extracts for Rapid Identification of Bioactive Compounds

Essentials of Botanical Extraction. http://dx.doi.org/10.1016/B978-0-12-802325-9.00010-0

10.1 INTRODUCTION

Drug discovery programs from botanicals endeavor to look out for bioactives possessing some kind of biological activity. However, the emergence of some kind of known and undesirable components with no chemical or pharmacological interest becomes almost inevitable. The process of identifying known compounds responsible for the activity of an extract prior to any isolation is referred to as dereplication. One of the main aims of dereplication procedures is to identify bioactives to an extent that will successfully conclude the extraction process efficiently. Moreover, it also provides information on how to progress through postextraction processes. Identification in such cases normally relies on the comparison of the extracted bioactives with standard marker compounds. Dereplication serves in streamlining the natural product drug discovery process. There are still a large proportion of natural products waiting to get discovered and, in the process, we do need methods and strategies for bypassing those bioactives that are already known. Analysis of data arising from a large set of samples can aid in the identification of useful structural scaffolds, which in turn can facilitate analog identification and prediction of drug-like property or drug likeliness.

In order to carry out dereplication, a number of general techniques are usually employed to gain some information about the compound that can be compared with other compounds or reference standards. Overall, a dereplication process involves the separation of a single metabolite by chromatographic methods, identification of compound(s) by spectroscopic methods, bioassays for evaluation of the biological activity, and database search for verification of the novelty of compound(s).

Usually, the dereplication process involves a series questions at each stage of the extraction based on the characteristics of the compound that will allow it to be assigned to a known compound, a general class of compounds, or an unknown compound. While performing the dereplication study, no single piece of information is conclusive on its own, but through partial identification, an overall idea is gained about the compound in hand.

10.2 TECHNIQUES ROUTINELY EMPLOYED IN DEREPLICATION STUDY

10.2.1 Separation Techniques

10.2.1.1 Use of Solvent Partitioning

During the initial extraction and subsequent concentration stages, the physical characteristics of unknown compounds are determined first. Simple solvent partitioning provides some information about the polarity and ionizability of the biologically active compounds within an extract, and this bit of information influences the choice of chromatographic separation methods later on. However, it should be kept in mind that the solvent partition technique does not result in compound identification, but does assist in the exclusion of classes to which the active principles cannot belong.

10.2.1.2 Thin Layer Chromatography, High-Performance Thin Layer Chromatography, and High-Performance Liquid Chromatography

If the compound under study is perceived to be a known compound that is available as a standard, then various available chromatographic techniques such as thin layer chromatography (TLC), high-performance thin layer chromatography (HPTLC), and high-performance liquid chromatography (HPLC) can be employed. If the compound under study shows different t_R (retention time for HPLC) and R_f (retention factor for TLC) values on an identical chromatographic system, then they cannot be the same compound; on the other hand, two samples of the same characteristics will have the same t_R and R_f values. Care should be taken in ensuring the degree of certainty while identifying the sample and the standard. In order to achieve the same, both should be injected together (coinjection/spiking/over spiking). On the chromatogram, if more than one peak is seen, then the two compounds are certainly different. Chromatographic behavior also provides some useful information about the nature of the compound. The presence of fatty acids in the column often results in a long retention time in an acidic medium while using a reverse phase column.

10.2.1.2.1 TLC Overlay Bioassay

When TLC is combined directly with a bioassay, some more information about an unknown compound can be gathered. This type of technique can be readily used to screen antimicrobial compounds. After development, the plates are overlaid with a thin layer of agar containing the test organism against which the compound under study is perceived to possess some activity. After a proper incubation period, zones of growth of inhibition in the agar can be seen in the regions of the TLC plate containing the active compound. By using such techniques, an unknown compound can be characterized within a mixture against a standard compound without the requirement for preparative isolation.

10.2.1.2.2 Bioautographic Methods for Effect-Directed Screening of Bioactives

Effect-directed bioautographic screening is an important tool in the search for new bioactive compounds from medicinal plants and other sources. It is also an additional detection mode for unknown constituents of samples. Simplified sample preparation steps and production of instant results are additional benefits of this technique.

Thin layer chromatography is frequently combined with other bioassays and is generally based on the inhibition/stimulation of growth or activity of test organisms like bacteria, yeast cells, mold spores, cell organelles, for example, chloroplasts or enzymes.

The enzyme inhibition test seems to be the most common among the above-mentioned methods and allows the detection and quantitative analysis of toxic substances in water, soil, air, and food samples. The HPTLC or TLC plate is sprayed or dipped sequentially in an enzyme solution and substrate (sometimes also dye) to give spots different in color from that of the background. The most popular enzymatic assay is the acetylcholinesterase inhibition test. Other enzymatic tests use glucosidase or xanthine oxidase inhibition.

Antioxidant and radical scavenging activities can be tested using β-carotene, DPPH (2,2-diphenyl-1-picrylhydrazyl) or ABTS (2,2′-azino-bis(3-ethylbenzthiazoline-6-sulfonic acid)) reagents. When the HPTLC or TLC plate is sprayed with β-carotene solution, orange zones on a cream–white background indicate the presence of antioxidants. In the case of DPPH, yellow spots on a purple background are observed in place of radical scavengers. The DPPH reagent can be replaced by ABTS, which gives pink or colorless spots on a green background. Similar to usual bioautography, both TLC–bioluminescence and the above-described methods can be merged with mass spectrometry (MS), infrared, or nuclear magnetic resonance (NMR) techniques to obtain constructive information about the structure and bioactivity of the compounds under study.

The bioassay, together with spectroscopic methods, provides full information about both the bioactivity and the structure of the analytes.

10.2.2 Spectroscopic and Hyphenated Techniques

Different types of spectroscopic techniques are generally used to carry out the partial or full identification of bioactive compounds that have been purified, but many spectroscopic methods can also be used at an earlier stage for compound identification within mixtures.

Hyphenated techniques have also played a major role in the identification of natural products and other organic compounds. These techniques combine a separation method (i.e., gas chromatography, GC; liquid chromatography, LC) with a structural identification technique (i.e., MS; ultraviolet–visible spectroscopy, UV–vis; NMR). Although a number of approaches dealing with the dereplication of natural products have been based on GC–MS, LC–UV (liquid chromatography UV spectroscopy), and LC–NMR (liquid chromatography NMR), LC–MS has been the most widely employed technique for this purpose.

The aim is to "eliminate" an unwanted compound, to assign it to a particular class or group, or to detect the presence/absence of a particular chemical group within a mixture.

10.2.2.1 Dereplication Based on UV Data

This technique partially identifies organic compounds through nondestructive physical characterization. Robert Burns Woodward and Louis Fieser made empirically derived rules that attempt to predict the wavelength of the absorption maximum (λ_{max}) in a UV–vis spectrum of a given compound. They were able to predict the λ_{max} for a number of conjugated systems, such as variously substituted enones and dienes, based on the available literature. Consequently, the UV spectrum once again became one of the most readily accessible means for structure elucidation, and interest has revived in exploiting its usefulness.

The advent of HPLC and UV diode array detectors has enabled the acquisition of a UV spectrum for every component represented in an HPLC chromatogram.

As all HPLC runs are carried out using the same solvent, under the same gradient profiles, the retention times as well as the UV profiles of unknowns can be compared against the library, and in this way, unknowns can be identified by "UV/retention time correlation study."

Today, most of the natural product chemists have analyzed hundreds of purified natural products by HPLC–UV, and consequently, many libraries of UV spectra with associated chromatographic behavior have been generated.

By comparing the UV spectra and chromatographic retention times of unidentified components obtained from the HPLC–UV analysis of crude extracts, it is possible to obtain a rapid indication of the presence of known compounds.

10.2.2.2 LC–MS and LC–MS–MS

MS is the most sensitive method for obtaining dereplication-related information about an unknown compound. Even microgram amounts are usually enough for several MS experiments, even though <1% of any sample undergoes ion formation by any single ionization technique.

LC–MS has played a valuable role in the screening of natural products as part of new drug discovery programs, facilitating rapid confirmation and dereplication of known and active compounds, and rapid discovery of new compounds.

MS depends on the determination of a mass-to-charge ratio (m/z), so, by definition, MS is only useful for characterizing molecules that can be ionized to a positive or negative charged state under controlled conditions. Currently, screening for natural products by LC–MS is accomplished primarily on the basis of molecular ion determination and product ion scanning using a quadrupole or ion trap mass spectrometer (Figure 10.1).

For dereplication and structure determination purposes, one of the most useful results from an LC–MS study is the identification of a compound's molecular ion. From the molecular ion, it is usually possible to determine the compound's molecular weight to the nearest atomic mass unit. This can be carried out by experimentally changing the mobile phase buffer and observing the resulting adduct ions in order to determine the adduct composition of the molecular ion, for example, $M^+ +H$, $M^+ +Li$, $M^+ +Na$, and $M^- -H$. The unit-molecular-weight information for an unknown compound can be used to dramatically reduce the number of possible structures under consideration for structure elucidation.

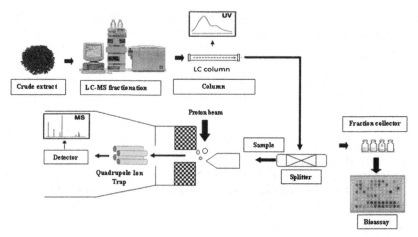

FIGURE 10.1 Liquid chromatography (LC)–mass spectrometry (MS) dereplication.

In the mass spectrometer, the molecular ion exists in a metastable high energy state, and fragmentation can usually be induced. This is most commonly carried out by Electron Ionization Mass Spectrometry or Laser desorption Mass Spectrometry or by further subjecting each ion to another round of ionization in a process of MS–MS. The pattern of fragmentation is reproducible, and if sufficient ions are detected, the profile can give nearly unique characteristic information.

Disadvantage with LC–MS is that the response is strongly dependent on the nature of the compounds to be analyzed, the solvent and buffer used for the separation, the flow rate, and the type of interface used. As a crude plant extract represents a complex mixture of metabolites, having various physicochemical properties, it will be difficult to find LC–MS conditions that are optimum for the ionization of all constituents. Often, it will be necessary to analyze the extract under different ionization conditions. On the other hand, the specific detection of given constituents can be performed at very low detection limits, provided that the correct ionization method is used. In a nut shell, it can be said that LC–MS can be an extremely powerful technique for screening crude plant extracts, but the right ionization conditions will have to be carefully optimized.

10.2.2.3 Liquid Chromatography–Nuclear Magnetic Resonance

NMR is a spectroscopic technique that by far provides the most useful information for the identification of natural products. Recently, advances in NMR spectroscopy have allowed HPLC to be practically interfaced directly with NMR. These advances include the use of higher field magnets ($\geq 500\,\text{MHz}$) and digital signal processing that have helped to address the issue of lack of sensitivity of this technique. NMR spectral data provide a great deal of structural information about a compound of interest. The NMR signal for each proton in a molecule provides structural information about the environment and the coupling partners of that proton. Therefore, NMR can easily discern structural differences between compounds of the same molecular weight (isobars) or even the same molecular formula (isomers).

NMR spectral data obtained using LC–NMR provide structural information, which other methods cannot. This provides a useful complement to the more sensitive and higher throughput methods of LC–MS and LC–DAD. Due to the limits in sensitivity and the lack of searchable NMR databases, LC–NMR is not yet ready for use as a frontline dereplication technique.

Disadvantages of LC–NMR are the difficulty in observing analyte resonances in the presence of the much larger resonances of the mobile phase. This problem has even worsened in the case of typical LC-reversed phase-operating conditions, where more than one protonated solvent was used and where the resonances changed frequencies during the analysis in the gradient mode. Further, the continuous flow of the sample in the detector coil complicated solvent suppression. These problems have now been overcome, thanks to the development of fast, reliable, and powerful solvent suppression techniques such as Water suppression Enhanced through T1 effects, which produced high-quality spectra in both on-flow and stop-flow modes.

10.2.2.4 Gas Chromatography–Mass Spectrometry

GC–MS is a hyphenated technique developed from the coupling of GC and MS. Mass spectra acquired by this hyphenated technique offer more structure-related information based on the interpretation of fragmentations of the ions. The fragment ions with different relative abundances can be compared with the library spectra. Nowadays, GC–MS is integrated with various online MS databases for several reference compounds with search capabilities that could be useful for spectra match for the identification of separated components. Compounds that are adequately volatile, small, and thermostable in GC conditions can be easily analyzed by GC–MS. Sometimes, polar compounds, with a number of hydroxyl groups, need to be derivatized prior to any analysis. The most common derivatization technique is the conversion of the analyte to its trimethylsilyl derivative.

Owing to its limited use in natural product studies, GC is not generally used for dereplication studies. It is actually suitable for volatile and nonpolar compounds.

10.2.3 Database Searching

Commercially available databases are available to assist in the dereplication process and often reduce the time taken for structure elucidation of known compounds.

- The Dictionary of Marine Natural Products (available online) containing >30,000 compounds.
- MarinLit contains data on marine organisms with a number of references from 1200 journals/books and data for approximately 21,000 compounds.

- AntiMarin—In this database, the number of methyl groups; the number of sp3-hybridized methylene, alkene, ether, and formyl groups can be searched.
- NAPRALERTTM—This contains the database of all natural products encompassing ethnomedical information, pharmacological/biochemical information of extracts of organisms *in vitro*, in situ, *in vivo*, in humans (case reports, nonclinical trials), and clinical studies.

Access to scientific databases is a fundamental and crucial step in a natural product search program. A thorough and extensive literature search is necessary where the following questions need to be addressed:

- Are there any previous literature reports known on the target organism under study?
- Is there any probability to isolate any novel compound(s)?
- What classes of compounds have been isolated from the species or the genus earlier?
- Are there any incomplete or poor NMR spectroscopic data for previously uncharacterized natural products?
- Has any new biological property for the known compound been overlooked?

It is of prime importance to tackle these questions during the early stages of the drug discovery process as one of the most common issues that occur is the time-consuming nature of isolation techniques, purifying, and determining the structure of novel bioactives.

10.3 STAGES WHERE DEREPLICATION IS APPLIED DURING TRADITIONAL AND MODERN APPROACHES OF DRUG DISCOVERY

10.3.1 Traditional Approach

The traditional approach in natural product drug discovery generally follows the path depicted in Figure 10.2.

Generally, while following the traditional approach in drug discovery, the concept of dereplication is not followed. However, the complicated nature of natural extracts often hampers the drug discovery process. Moreover, the limited quantity of the starting material and the final identified products hinder the entire procedure, which is significantly slow, laborious, and expensive. Moreover, it suffers from reproducibility- and sensitivity-related issues.

10.3.2 Bioassay-Guided Isolation Approach

An important improvement of this approach with respect to the previous one comprises the so-called "bioassay-guided isolation," which introduces a more rational and targeted concept throughout the classical procedure as shown in Figure 10.3.

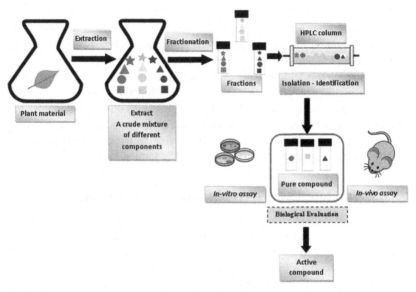

FIGURE 10.2   Traditional approach in natural product drug discovery. HPLC, high-performance liquid chromatography.

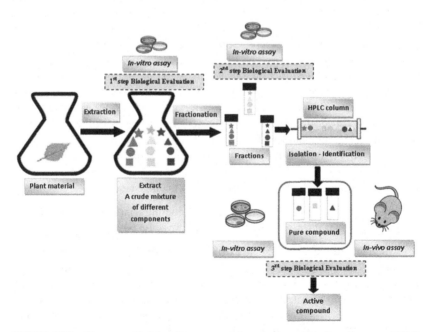

FIGURE 10.3   Bioassay-guided isolation approach in natural product drug discovery. HPLC, high-performance liquid chromatography.

By the successive biological evaluation on the obtained extracts, derived fractions, and final purified compounds, the entire process is monitored and guided based on bioactivity, thereby focusing more selectively on the discovery of bioactives.

Even though this approach is much more focused on biological targets, there are some specific restrictions that limit its use. Competition and synergism phenomena of constituents present in extracts and fractions often result in false positive and negative results, thus jeopardizing the entire mission of drug discovery. Thus, a methodology has to be implemented in order to prevent the occurrence of false negatives and rediscovery of the already known bioactives.

10.3.3 Structure-Based Dereplication-Driven Strategy

This approach is more structure oriented and uses a higher degree of powerful analytical and elucidation techniques as well as dereplication strategies that result in the rapid identification of already known molecules. The identification of a secondary metabolite early in the discovery process (e.g., from extracts or fractions) significantly increases the entire progression of new actives discovery eliminating the repeated and laborious isolation stages. In modern days, employment of several spectral databases, informatics, intelligent correlation software, and hyphenated analytical apparatuses enables the prioritization of extracts/fractions leading to the active compound(s) isolation selectively.

Overall, the above-mentioned work flows are followed by the majority of phytochemistry research groups either alone or as combinations. However, besides the general discovery concepts, significant progress has been made in the individual steps of extraction, separation, isolation, and identification as well as dereplication. Technological advanced instrumentation, bioactives enrichment capabilities, affinity-based screening formats, and devices are increasingly incorporated for accelerating not only the detection and purification of an active secondary metabolite but also for activity-dereplication purposes.

Figure 10.4 highlights the stages where dereplication is applied in the drug discovery process.

Dereplication can be applied during the following stages:

1. After primary extraction and just before initiating a large-scale extraction in order to gather information on how to progress the postextraction steps.

After performing a small-scale extraction, the primary extracts thus obtained are actually an extract library of structurally diverse secondary metabolites that can be classified. Thus, the natural product source exhibiting the highest chemical diversity can be selected for large-scale extraction followed by isolation and purification. Modern hyphenated techniques such as LC–MS, LC–MS/MS, and LC–NMR have been built for the rapid analysis and dereplication of complex mixtures obtained from nature. Apart from this chemical characterization,

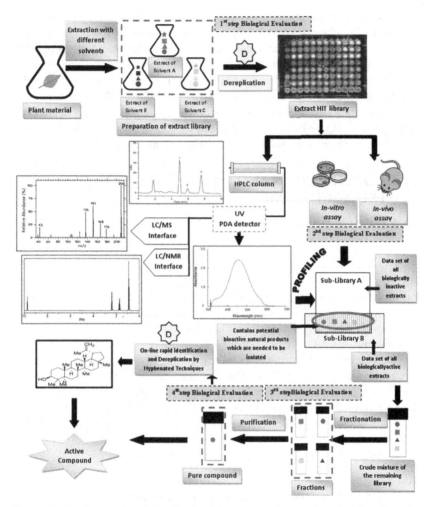

FIGURE 10.4 Structure-based dereplication-driven strategy in natural product drug discovery. HPLC, high-performance liquid chromatography; LC, liquid chromatography; MS, mass spectrometry; NMR, nuclear magnetic resonance.

bioactivity profiles can be generated using microfractionation of crude extracts employing 96-well plates. These preliminary findings from a crude extract library give a very good biochemical profiling of Hits (compounds with desirable characteristics) within crude extracts, thus generating Extract Hit Libraries.

2. During bioactivity guided fractionation to obtain lead compounds for subsequent lead isolation.
3. The next step is to determine whether the leads that actually did not match with the internal database libraries are in fact novel or not.

Internal database libraries are constructed by using some basic information or criteria like taxonomy of the organism, bioactivity, molecular formula, molecular weight, and spectroscopic data. Just before isolation, only a very few criteria are known, and as a result, a long list of potential matches are generated while going for database delving, and the main challenge is to narrow down that list. Applications of MS that rely on molecular weight and formula can be one of the most powerful means to narrow down this search. Various hyphenated techniques like LC–MS, LC–MS/MS, and LC–NMR have been applied as an effective tool prior to any bioassay-guided isolation.

10.4 CONSTRUCTION AND CHARACTERIZATION OF EXTRACT LIBRARIES

10.4.1 Crude Extract Library

As a starting point for the generation of a crude extract library, the first step is the collection of organisms (biota). Once collected, the biota requires processing in such a way so as to create an extract suitable for both biological profiling and chemical profiling or fingerprinting. Biological profiling involves testing the extracts in assays (e.g., antibacterial, antifungal). Biological screening is more sensitive than chemical profiling and can detect minor compounds.

Chemical profiling of the extracts is generated by liquid chromatography–diode array detector–evaporative light scattering detector and is analyzed for the presence of a secondary metabolite. This type of profiling actually aids in the dereplication of known compounds or leads and thus can be applied to lower repetition in the overall library. A schematic representation of a crude library is shown in Figure 10.5.

Advantages of a crude extract library are as follows:

1. Can be prepared with minimum processing steps.
2. Requires less resource for subsequent sample preparation steps.
3. Has a high degree of diversity.

Disadvantages of a crude extract library are as follows:

1. Presence of minor secondary metabolites may go unnoticed and undetected.
2. Massive and extensive scale-up is required to identify the active constituents.
3. Chemical nature of the component present often presents challenges. Moreover, the presence of highly polar or lipophilic compounds and the crude extracts can interfere in the proper functioning of several bioassays, resulting in false positives.
4. Rediscovery of known structures may occur.
5. Chemically unattractive compounds may be isolated.
6. Time and resource intensive follow-up is required.

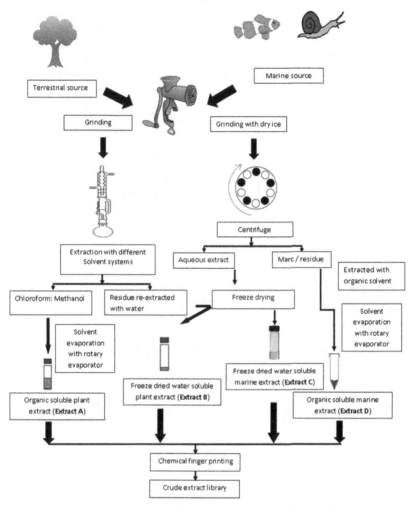

FIGURE 10.5 Schematic representation of a crude library.

10.4.2 Prefractionated (Semipure) Library

Crude extract libraries owing to their inherent complexity face several hurdles when they are introduced into the drug discovery phase employing highthroughput screening (HTS) techniques. Once a "HIT" gets identified from an extract library, the heterogeneity of the library samples adds additional levels of complexity. Extracts meeting the criteria and showing potential for further fractionation are concentrated and divided equally in two portions as shown in Figure 10.6. One portion is stored at −20 °C, and the remaining is used for prefractionation (fractionation of crude extract prior to biological testing) and subsequent plating. Each fraction may vary in complexity from a mixture of many leads to a single

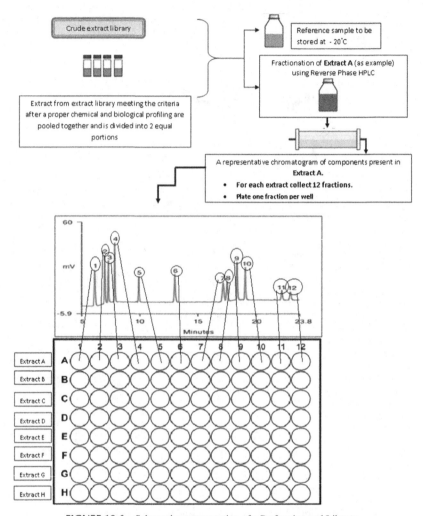

FIGURE 10.6 Schematic representation of a Prefractionated Library.

major compound of high purity. Fractionation through various conventional separation techniques such as column chromatography, liquid-liquid extraction is being gradually replaced with SPE. It utilizes the affinity of solutes (dissolved in mobile phase) for a solid phase (i.e., the stationary phase) through which the sample solution is passed so that the sample mixture gets separated into desired and undesired components. By doing so, either the targeted analyte of choice or undesired compounds in the sample mixture are retained in the stationary phase. The portion of the sample mixture that passes through is either collected or thrown away depending on whether it contains the desired analyte or any impurities. If any desired analyte gets retained with the stationary phase, it can be removed after rinsing with an appropriate eluent that actually has more affinity toward the stationary phase than the analyte.

Other techniques include preparative reversed phase HPLC as it is reproducible, robust, automated, has a good resolution, and requires minimal sample preparation steps.

Advantages of a Prefractionated Library:

1. Samples generated through this procedure are simple mixtures and require few purification steps toward getting an active compound.
2. Extremely polar and nonpolar substances easily get separated.
3. Chances of discovering a novel biologically active metabolites increases.

Disadvantages of a Prefractionated Library:

1. As the prefractionation step increases the number of samples, the cost to maintain such a vast library increases as well.

10.4.3 Purified Natural Product Library

A purified natural product library will definitely be less chemically diverse compared to a library extract and can be handled like any other pure synthetic molecular libraries. Building a natural product library actually offers an alternative to extract libraries and helps to overcome the various shortcomings associated with extract libraries. A purified natural product library is formed after several fractionation steps leading to clear samples thereby increasing the chances for discovering of minor bioactive compounds.

A purified natural product library offers advantages in the faster detection of quality HITS as no subsequent isolation step is required.

Despite various challenges faced in plant-based drug discovery, bioactives isolated from plants will still remain an essential entity in the search for new therapeutics. Aptly utilizing these resources and tools in bioprospecting will definitely help in discovering novel bioactives from plants by using modern drug discovery techniques. But care should be taken for the continuous improvement in the dereplication process, compound isolation techniques, structure elucidation, along with a cautious approach in selecting drug targets for the screening of Natural Product libraries.

FURTHER READING

Michel, T., Halabalaki, M., Skaltsounis, A.-L., 2013. New concepts, experimental approaches, and dereplication strategies for the discovery of novel Phytoestrogens from natural sources. Planta Medica 79, 514–532.

Patel, K.N., Patel, J.K., Patel, M.P., Rajput, G.C., Patel, H.A., 2010. Introduction to hyphenated techniques and their applications in pharmacy. Pharmaceutical Methods 1 (1), 2–13.

Index

Note: Page numbers followed by "b", "f" and "t" indicate boxes, figures and tables respectively.

Printed in the United States
By Bookmasters